Capacitive Silicon Resonators

Performance Enhancement Methods

Capacitive Silicon Resonators

Resonators

Performance Enhancement Methods

Authored by
Nguyen Van Toan and Takahito Ono

CRC Press
Taylor & Francis Group
Boca Raton London New York

CRC Press is an imprint of the
Taylor & Francis Group, an **informa** business

CRC Press
Taylor & Francis Group
6000 Broken Sound Parkway NW, Suite 300
Boca Raton, FL 33487-2742

Printed on acid-free paper

International Standard Book Number-13 978-0-367-21776-1 (Hardback)

Library of Congress Cataloging-in-Publication Data

Names: Toan, Nguyen Van, author. | Ono, Takahito, author.
Title: Capacitive silicon resonators : performance enhancement methods /
Nguyen Van Toan and Takahito Ono.
Description: Boca Raton, FL : CRC Press/Taylor & Francis Group, [2020] |
Includes bibliographical references.
Identifiers: LCCN 2019010375| ISBN 9780367217761 (hardback : acid-free paper)
| ISBN 9780429266010 (ebook)
Subjects: LCSH: Dielectric resonators. | Electric resonators. |
Nanoelectromechanical systems. | Semiconductor devices.
Classification: LCC TK7872.D53 T63 2020 | DDC 621.3815--dc23
LC record available at https://lccn.loc.gov/2019010375

Visit the Taylor & Francis Web site at
http://www.taylorandfrancis.com

and the CRC Press Web site at
http://www.crcpress.com

Printed and bound in Great Britain by
TJ International Ltd, Padstow, Cornwall

Contents

PART 2A   *Performance Enhancement Methods for Capacitive Silicon Resonators: Fabrication Technologies*

PART 2B   *Performance Enhancement Methods for Capacitive Silicon Resonators: Design Considerations*

Preface

Microfabricated resonators play an essential role in a variety of applications, including mass sensing, timing reference applications, and filtering applications. Many transduction mechanisms, including piezoelectric, piezoresistive, and capacitive mechanisms, have been studied to induce and detect the motion of resonators. All the aforementioned methods possess both advantages and disadvantages. In this work, we focus on a capacitive sensing method based on the measurement of the change in capacitance between a sensing electrode and the resonant body. Capacitive resonators are typically employed for sensing and timing applications. Capacitive resonators exhibit a high quality (Q) factor, which results in stability and low internal friction. However, their drawbacks are their large motional resistances and high insertion losses, which make it difficult to satisfy oscillation conditions. Also, a large motional resistance results in significant phase noise in oscillators; therefore, the motional resistance should be as small as possible.

In this book, we review the recent progress in performance enhancement methods for capacitive silicon resonators, which are mainly based on our works. Several technological approaches including hermetic packaging based on a low temperature co-fired ceramic (LTCC) substrate, deep reactive ion etching, neutral beam etching technology, and metal-assisted chemical etching, as well as design considerations for a mechanically coupled resonator, a selective vibration of high-order modes, movable electrode structures, and piezoresistive heat engines were investigated to achieve a small motional resistance, a low insertion loss, and a high quality factor.

There are a number of excellent comprehensive books on microfabricated resonators. This book is meant to introduce and suggest several technological approaches together with design considerations for the performance enhancement of capacitive silicon resonators. Contents of this book are based on experimental research done at Tohoku University, Sendai, Japan. We hope that this book may be a useful reference to those working in the field of micro/nanotechnology.

Preface

Acknowledgments

We would like to express our greatest gratitude to Prof. Nguyen Van Hieu, Prof. Masaya Toda, Prof. Naoki Inomata, Prof. Hidetoshi Miyashita, Prof. Yusuke Kawai, Prof. Shuji Tanaka, Prof. Hiroki Kuwano, Prof. Kazuhiro Hane, Prof. Seiji Samukawa, Prof. Yu-Ching Lin, Prof. Tomohiro Kubota, Prof. Yunheub Song, Dr. Young-Min Lee, Dr. Yong-Jun Seo, Dr. Zhonglie An, Dr. Yacine Belbachir Remi, Dr. Gaopeng Xue, Dr. Zhuqing Wang, Dr. Mahabub Hossain, Dr. Jinhua Li, Dr. Nguyen Huu Trung, Dr. Le Van Minh, Dr. Ho Thanh Huy, Dr. Nguyen Chi Nhan, Dr. Vu The Dang, Mr. Nguyen Van Nha, Mr. Tsuyoshi Shimazaki, Mr. WenShuo Zhang, Mr. Dong Zhao, Ms. Xiaoyue Wang and Ms. Truong Thi Kim Tuoi for their collaboration and contribution in this book.

We would like to express our sincere appreciation to Nora Konopka, Prof. Ioana Voiculescu, Prof. He Liang, and Dr. Halubai Sekhar, who were willing to review this book.

Our thanks are to all the members of Ono/Toda laboratory, past and present, for creating a great working environments and giving much help. We are grateful to the secretaries – Ms. Fumie Takei, Ms. Tomomi Kuzuno, Ms. Kanako Sasaki, Ms. Megumi Nishimura, and Ms. Soko Matsuda – for assisting in many different ways.

A part of this work was performed in the Micro/Nanomachining Research and Education Center, the Nishizawa center, and WPI-AIMR of Tohoku University. This work was supported in part by the Council for Science, Technology and Innovation (CSTI), Cross-ministerial Strategic Innovation Promotion Program (SIP), and also supported in part by JSPS KAKENHI for Young Scientists B (Grant number: 17K14095).

About the Authors

Nguyen Van Toan received his BS degree in 2006 and his MS degree in 2009 in physics and electronics, respectively, from the University of Science, Vietnam National University, Ho Chi Minh City, Vietnam. He received his Doctor of Engineering from Tohoku University in 2014 for research on silicon capable of utilizing large-scale integration (LSI) for application to timing devices. He works as an assistant professor in the Department of Mechanical Engineering, Graduate School of Engineering at Tohoku University. His current research interests include capacitive silicon resonators, optical modulator devices, capacitive micromachined ultrasonic transducers, thermal electric power generators, Knudsen pump, ion transportation, and metal-assisted chemical etching.

Takahito Ono is currently a professor at Mechanical Systems Engineering, Graduate School of Engineering in Tohoku University. He was born in Hokkaido, Japan, on 12 July 1967. He received the BS degree in physics from Hirosaki University, Japan, in 1990 and an MS degree in physics from Tohoku University, Japan. He received a Doctor of Engineering in mechatronics and precision engineering from Tohoku University in 1996. From 1996 to 2001, he was a research associate and lecturer in the Department of Mechatronics and Precision Engineering, Tohoku University. He studied nanomachining and the scanning probe and its related technologies including high-density storage devices. Between 2001 and 2009, he was an associate professor and developed nanomechanics and nanomechanical sensors. Since 2009, he has been a professor at Tohoku University. His expertise is in the areas of microelectromechanical systems (MEMS), nanoelectromechanical systems (NEMS), silicon-based nanofabrication, and ultrasensitive sensing based on NEMS/MEMS. Also during 2012–2014 he was director of the Micro/Nanomachining Research and Education Center, Tohoku University. Since 2017 he has served as a director of the Micro System Integration Center (μSiC), Tohoku University. During 2013–2016, he has held the additional post of Professor of Guest Courses, Mechanical Departments, University of Tokyo, and is working on nanomechanics. He is editor in chief of *IEEJ Transactions on Sensors and Micromachines*, editor member of *Microsystems & Nanoengineering*, and advisory board member of *Journal of Micromechanics and Microengineering*. His expertise is in the area of microelectromechnical systems (MEMSs), nanoelectromechanical systems (NEMSs), silicon-based nanofabrication, and ultra-sensitive sensing based on resonating device. Recent interests cover nanomaterials and their process integration, biomedical sensors, and micro energy.

1 Introduction

1.1 INTRODUCTION

A resonator is a structure that exhibits resonant behavior, meaning a resonator oscillates naturally with greater amplitude at its resonant frequencies than at others. The oscillations in a resonator can be either electromagnetic or mechanical. Guitar strings resonating at frequencies in audible range between 20 Hz and 20 kHz is a typical example of mechanical resonator in macro scale. A micromechanical resonator is a micromachined mechanical structure of much smaller size in a comparison with a guitar that can operate at much higher frequencies from kilohertz up to megahertz or even gigahertz. This frequency range is interesting for a variety of applications in electronic circuits and systems, which has attracted more attention and investment in recent years. Many Internet of things (IoT) devices can be connected to the Internet via wireless networks. The increasing amount of transmitted and received information requires accurate data transmissions. Satisfying these issues, micro clock generators for transmitters and receivers with a smaller size and higher performance are required.

In an electric circuit, the micromachined mechanical resonator converts electrical energy to mechanical energy and then reverses this process to gain electrical signals as output. The electromechanical transducer in the resonator implements the energy conversion process, so thus the structure is known as a microelectromechanical resonator (microfabricated resonator) and the whole system belongs to the concept of microelectromechanical systems (MEMS). With MEMS it is generally accepted that at least one of their dimensions is in the range from hundreds of nanometers to hundreds of micrometers. MEMS were conceived under the integrated circuit (IC) paradigm, in which silicon is the base material for device fabrication. Therefore, MEMS share the most technologies and know-how from the mature IC industry. Along with the miniaturization of electric elements, MEMS and the complementary metal–oxide semiconductor (CMOS) circuit on one chip became dominant for integrated applications, whose potential is not only exploited for ICs but also for a variety of MEMS-based applications.

1.2 REVIEW OF MICROFABRICATED RESONATORS

Microfabricated resonators play an essential role in a variety of applications [1–9]. Resonating ultrathin cantilevers for mass sensing have been presented [1, 2], in which the resonant frequency of the cantilever monitors the adsorbed mass on its surface. Timing reference applications that use resonator structures to generate clock signals in electronic systems have been demonstrated in many studies [3–6]. Microfabricated resonators are also employed for filtering applications [7–9] that can be of use in radio frequency transmitters and receiver modules. Many transduction mechanisms, consisting of piezoelectric [7–10], thermal-piezoresistive [11, 12], and capacitive [5, 6, 13] mechanisms have been studied to induce and detect the motion of resonators. All the aforementioned methods possess both advantages and disadvantages. Figure 1.1 illustrates microfabricated resonator structures including a piezoelectric resonator, a thermal-piezoresistive resonator, and a capacitive resonator.

Piezoelectric resonators (Figure 1.1a), usually used for filtering applications [9, 14], offer low insertion loss and small motional resistance, but the quality factors (Q) of resonators are small and good piezoelectric materials are vital. Piezoelectric micromechanical resonators have similar structures to quartz structures that consist of piezoelectric material sandwiched between two metal electrodes. When the electric polarization applies to these electrodes, mechanical stress occurs or vice versa. Many piezoelectric materials, such as lead zirconate titanate (PZT), aluminum nitride (AlN), or zinc oxide (ZnO), have been investigated; however, further improvements are necessary to achieve high piezoelectric coefficients and low residual stress.

Thermal-piezoresistive resonators (Figure 1.1b) were presented [11, 15] for the applications of oscillators and thermal-piezoresistive energy pumps. Piezoresistive sensing with the thermal actuation shows a high quality factor and low insertion loss; however, it faces significant power consumption. Also, this technique is not appropriate for the construction of high-sensitivity products due to strong effects from environmental conditions, such as temperature dependence. Recently, there have been some efforts to improve the piezoresistive transduction sensitivity [16, 17]. Self-sustained oscillators without any amplifying circuitry can be achieved by piezo-resistive heat engines, which are based on thermodynamic cycles. Their operation principles are as follows: due to the higher resistance of narrow piezoresistive beams over other parts of the resonator structure, the piezoresistive beams are heated when a direct current (DC) voltage is applied to the piezoresistive elements. This results in the expansion of the piezoresistive beams, which causes the resistance to increase due to the piezoresistive effect. As a consequence of this resistance increase, the current passing through the piezoresistive beams decreases, which also decreases the temperature of the piezoresistive beam. Thus, the piezoresistive beams are compressed, and the resistance decreases due to the piezoresistive effect, resulting in an increase in the current. From this, thermomechanical actuation power is generated

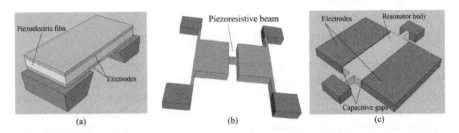

FIGURE 1.1   Microfabricated resonator structures: (a) piezoelectric resonator, (b) thermal-piezoresistive resonator, (c) capacitive resonator.

by the aforementioned cycle in the piezoresistive elements. Therefore, self-oscillation can be achieved [16].

Capacitive resonators (Figure 1.1c), typically employed for sensing and timing applications, are based on the measurement of the change in capacitance between a sensing electrode and the resonant body. Capacitive resonators exhibit a high Q factor, which results in the stability and low internal friction. However, their drawbacks are large motional resistances and high insertion losses that make it difficult to satisfy oscillation conditions. Also, the large motional resistance results in significant phase noise in the oscillator; therefore, the motional resistance should be as small as possible. Methods for lowering the motional resistance [18–25] include reducing the capacitive gap width [18–20], and increasing the overlap area of the capacitance, the quality factor Q, and the polarization voltage V_{DC}. Unfortunately, each of these methods has some drawbacks. Decreasing the capacitive gap width is very efficient in lowering the motional resistance; however, fabrication of a nanogap is complicated, and the applicable maximum polarization voltage decreases due to the pull-in phenomenon. Smaller capacitive gaps result in lower pull-in voltages that easily cause short-circuit situations. An increase in the electrode areas being utilized, such as resonator arrays [4, 21] and mechanical coupling [22], can reduce the motional resistance, but the frequency response and Q factor of the resonators suffer from the mismatches in the individual resonant frequencies. To increase the Q factor, some of the methods to reduce the energy losses, including external and internal losses, have been reported, but there are material and structural limitations [5, 23–25]. Increasing the polarization voltage is limited because high voltage is not available on a CMOS chip. It means that more challenges and difficulties face the large-scale integration (LSI) process for silicon resonators.

Based mainly on our experiments, in this book we review the recent progress of capacitive silicon resonators, including different fabrication technologies and design considerations to achieve small motional resistance, low insertion loss, and high Q factor. The technological solutions and design considerations presented in this book can be effective, not only for capacitive silicon resonators but also for other fields of micro- and nanosystems.

1.3   STRUCTURE OF THE BOOK

This book is a comprehensive scientific monograph about the fabrication technologies and design considerations of capacitive silicon resonators. It contains 12 chapters with about 100 image illustrations and over 200 references.

Chapter 1 introduces the motivation and structure of the book.

Chapter 2, "Capacitive Silicon Resonator Structures," presents an overview of device structures and working principles, an equivalent circuit model, a finite element model, and key parameters of capacitive silicon resonators.

Chapter 3, "Fabrication Techniques for Capacitive Silicon Resonators," briefly introduces some fundamental fabrication techniques for capacitive silicon resonators.

Chapter 4, "Hermetically Packaged Capacitive Silicon Resonators on LTCC Substrate," presents the new approach for hermetically packaged silicon resonators.

The packaging process using LTCC substrate was successfully performed, and resonance with a high Q factor was observed. The proposed hermetically sealed resonators based on the LTCC substrate can simplify LSI for timing devices.

Chapter 5, "A Long-Bar-Type Capacitive Silicon Resonator with a High Quality Factor," demonstrates that the longer bar-type resonant body has a higher Q factor and a lower motional resistance than those of the shorter bar-type resonant body. The device was hermetically packaged using an LTCC substrate on the basis of anodic bonding technology, and the resonant characteristics before and after the packaging process were evaluated and compared.

Chapter 6, "Capacitive Silicon Resonators Using Neutral Beam Etching Technology," is a description and comparison of capacitive silicon resonators produced by both deep reactive ion etching (DRIE) and neutral beam etching (NBE). The devices fabricated by NBE provided a higher Q factor, a lower insertion loss, and a smaller motional resistance than those fabricated by DRIE.

Chapter 7, "Capacitive Silicon Resonators with Narrow Gaps Formed by Metal-Assisted Chemical Etching," presents a simple fabrication technology for producing narrow gaps using metal-assisted chemical etching (MACE). MACE is conducted by a simple operation with a low cost, which is performed in a wet etching solution. It enables the formation of anisotropic silicon structures at room temperature and atmospheric pressure. This proposed method shows a simple and economical way for fabricating capacitive silicon resonators.

Chapter 8, "Mechanically Coupled Capacitive Nanomechanical Silicon Resonators," reports on the mechanically coupled capacitive nanomechanical silicon resonator structure that can be used for increasing motional capacitance and lowering motional resistance for emerging sensitive applications and image and data processing technologies. Resonant peaks of the mechanically coupled capacitive nanomechanical silicon resonators were observed, which show that most nanomechanical resonators are mechanically coupled and synchronized.

Chapter 9, "Capacitive Silicon Nanomechanical Resonators with Selective Vibration of High-Order Mode," presents the high-order mode capacitive silicon resonators by placing the driving electrodes along the resonant body. It indicates that high-order mode vibration structures can achieve lower insertion loss and smaller motional resistance over those of low-order mode structures.

Chapter 10, "Capacitive Silicon Resonators with Movable Electrode Structures," is a comparison of the capacitive silicon resonator with and without movable electrode structures. The frequency response of the capacitive silicon resonator with movable electrode structures shows a smaller motional resistance, a lower insertion loss, and a wider tuning frequency range than that of the capacitive silicon resonator without movable electrode structures.

Chapter 11, "Capacitive Silicon Resonators with Piezoresistive Heat Engines," reports a performance enhancement method of capacitive silicon resonators by piezoresistive heat engines. Capacitive silicon resonators with single and multiple piezoresistive beams were fabricated and evaluated. An improvement in insertion loss and reduction in motional resistance were achieved.

Finally, Chapter 12 presents the conclusions of the book.

REFERENCES

1. Ono, T., Esashi, M., Mass sensing with resonating ultra-thin silicon beams detected by a double-beam laser Doppler vibrometer, *Measurement Science Technology*, **15**, 1977–1981, 2004.
2. Kim, S.J., Ono, T., Esashi, M., Mass detection using capacitive resonant silicon resonator employing LC resonant circuit technique, *Review of Scientific Instruments*, **78**, 085103, 2007.
3. Nguyen, C.T.C., MEMS technology for timing and frequency control, *IEEE Transactions on Ultrasonics, Ferroelectronics, and Frequency Control*, **54**, 251–270, 2007.
4. Ayazi, F., MEMS for integrated timing and spectral processing, *IEEE Custom Integrated Circuits Conference*, Rome, 65–72, 2009.
5. van Beek, J.T.M., Puers, R., A review of MEMS oscillators for frequency reference and timing applications, *Journal of Micromechanics and Microengineering*, **22**, 013001, 2012.
6. Toan, N.V., Miyashita, H., Toda, M., Kawai, Y., Ono, T., Fabrication of an hermetically packaged silicon resonator on LTCC substrate, *Microsystem Technologies*, **19**, 1165–1175, 2013.
7. Piazza, G., Stephanou, P.J., Porter, J.M., Wijesundara, M.B.J., Pisano, A.P., Low motional resistance ring-shaped contour-mode aluminum nitride piezoelectric micromechanical resonators for UHF application, *18th IEEE International Conference on Micro Electro Mechanical Systems*, Miami Beach, FL, 20–23, 2005.
8. Sorenson, L., Fu, J.L., Ayazi, F., One-dimensional linear acoustic bandgap structures for performance enhancement of AlN-on-silicon micromechanical resonators, *16th International Conference on Solid State Sensors, Actuators and Microsystems*, Beijing, China, 918–921, 2011.
9. Nguyen, N., Johannesen, A., Hanke, U., Design of high Q thin film bulk acoustic resonator using dual-mode reflection, *IEEE International Ultrasonics Symposium*, Chicago, IL, 487–490, 2014.
10. Ali, A., Lee, J.E.-Y., Novel platform for resonant sensing in liquid with fully electrical interface based on an in-plane-mode piezoelectric-on-silicon resonator, *Procedia Engineering*, **120**, 1217–1220, 2015.
11. van Beek, J.T.M., Steeneken, P.G., Giesbers, B., A 10 MHz piezoresistive MEMS resonator with high Q, *IEEE International Frequency Control Symposium and Exposition*, Miami, FL, 475–480, 2006.
12. Rahafrooz, A., Pourkamali, S., Thermal-piezoresistive energy pumps in micromechanical resonator structures, *IEEE Transactions on Electron Devices*, **59**, 3587–3593, 2012.
13. Ho, G.K., Sundaresan, K., Pourkamali, S., Ayazi, F., Low motional impedance highly tunable I2 resonators for temperature compensated reference oscillators, *18th IEEE International Conference on Micro Electro Mechanical Systems*, Miami Beach, FL, 116–120, 2005.
14. Hashimoto, K.Y., Kimura, T., Matsumura, T., Hirano, H., Kadota, M., Esashi, M., Tanaka, S., Moving tunable filters forward, *IEEE Microwave Magazine*, **16**, 89–97, 2015.
15. Rahafrooz, A., Pourkamali, S., Thermal-piezoresistive energy pumps in micromechanical resonator structures, *IEEE Transactions on Electron Devices*, **59**, 3587–3593, 2012.
16. Steeneken, P.G., Le Phan, K., Goossens, M.J., Koops, G.E.J., Brom, G.J.A.M., van der Avoort, C., van Beek, J.T.M., Piezoresistive heat engine and refrigerator, *Nature Physics*, **7**, 354–359, 2011.

17. Ramezany, A., Mahdavi, M., Pourkamali, S., Nanoelectromechanical resonant narrow band amplifiers, *Microsystems & Nanoengineering*, **2**, 16004, 2016.
18. Pourkamali, S., Ho, G.K., Ayazi, F., Low impedance VHF and UHF capacitive silicon bulk acoustic wave resonators – Part I: Concept and fabrication, *IEEE Transactions on Electron Devices*, **54**, 2017–2023, 2007.
19. Pourkamali, S., Ho, G.K., Ayazi, F., Low impedance VHF and UHF capacitive silicon bulk acoustic wave resonators – Part II: Measurement and characterization, *IEEE Transactions on Electron Devices*, **54**, 2024–2030, 2007.
20. Toan, N.V., Toda, M., Kawai, Y., Ono, T., A capacitive silicon resonator with a movable electrode structure for gap width reduction, *Journal of Micromechanics and Microengineering*, **24**, 025006, 2014.
21. Qishu, Q., Pourkamali, S., Ayazi, F., Capacitively coupled VHF silicon bulk acoustic wave filters, *IEEE Ultrasonics Symposium*, 1649–1652, 2007.
22. Toan, N.V., Shimazaki, T., Ono, T., Single and mechanically coupled capacitive silicon nanomechanical resonators, *Micro & Nano Letters*, **11**, 591–594, 2016.
23. Lee, J.E.Y., Yan, J., Seshia, A.A., Study of lateral mode SOI-MEMS resonators for reduced anchor loss, *Journal of Micromechanics and Microengineering*, **21**, 045011, 2011.
24. Toan, N.V., Kubota, T., Sekhar, H., Samukawa, S., Ono, T., Mechanical quality factor enhancement in silicon micromechanical resonator by low-damage process using neutral beam etching technology, *Journal of Micromechanics and Microengineering*, **24**, 085005, 2014.
25. Toan, N.V., Toda, M., Kawai, Y., Ono, T., A long bar type silicon resonator with a high quality factor, *IEEJ Transactions on Sensors and Micromachines*, **134**, 26–31, 2014.

Part 1

Backgrounds

2 Capacitive Silicon Resonator Structures

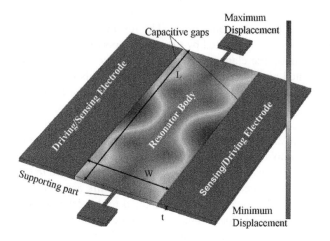

2.1 DEVICE STRUCTURE AND WORKING PRINCIPLE

A capacitive resonator is composed of a resonant body between driving/sensing electrodes. The resonant body, namely, the vibrating parts, is separated from stationary electrodes by narrow gaps that should be as small as possible. In general, resonant bodies are generated with dry plasma etching on the device layer of silicon on insulator (SOI) wafer, and released from the handling layer through undercutting the buried oxide layer. The types of capacitive resonators are classified according to the shape of the resonant bodies including beams [1, 2], bars [3–5], square plates [6, 7], and disks [8]. Depending on the shape, the resonant bodies are anchored with substrate at certain positions via supporting beams. For better understanding, a long, narrow resonant body "floating" on the substrate and anchored at two ends (clamped–clamped beam type) can be imagined as a section of a bridge over water. An actuating resonant body through electromechanical transduction is the same as if the bridge is blown by the wind with an external periodic frequency that coincides with its natural structural frequency. The bridge exhibits its own resonant behavior in macroscale by swinging, turning, or twisting in the largest amplitude. In the micro/nanoscale, resonators vibrate in various modes at different orders. Micro/nanomechanical resonators can be broadly classified according to their mode of operation, namely, flexural, torsional, and bulk mode devices. Depending on the shape of resonant bodies, resonators with the capacitive transduction mechanism operate in either flexural or bulk modes. Due to larger stiffness of resonator structures operating in bulk mode, such devices can demonstrate much higher frequencies compared to the flexural modes.

Figure 2.1a illustrates a two-arm-type capacitive resonator structure, also called a longitudinal beam resonator. This resonator is electrostatically actuated by applying a voltage to the electrodes at its ends. Two anchors support the middle point of the resonant body through symmetrical supporting beams. The deformation, in terms of changing the length of the resonant body, is unity and longitudinal, which results in vibration in the length extensional mode.

Figure 2.1b demonstrates a bar-type capacitive resonator structure that is excited in its horizontal width extensional mode. The resonant body is a bar shape, placed between two electrodes, supported by two thin supporting parts on the sides.

Figure 2.1c shows the clamped–clamped beam-type structure. The resonator is electrically excited and vibrated at flexural mode. In this mode, maximum deflection in the horizontal direction occurs at the center of the nanomechanical beam (Table 2.1).

The general schematic diagram of an electrostatically operated bar-type resonator is shown in Figure 2.2. To actuate the resonant body, a combination of an alternating voltage V_{AC} and a polarization voltage V_{DC} is conducted. The capacitive resonator's operation is described as follows: An AC voltage V_{AC} is supplied to the driving electrode, resulting in the electrostatic force inducing a bulk acoustic wave in the resonant body. Additional DC voltage V_{DC} is applied to the driving/sensing electrodes to amplify the electrostatic force. This electrostatic force makes the resonant body actuated. Small changes in the capacitive gap width between the sensing electrode and resonant body generate a voltage on the sensing electrode.

F_1 and F_2 in Figure 2.2 are the absolute electrostatic forces on driving and sensing sides, respectively, which can be calculated in Equations 2.1 and 2.2. A delta

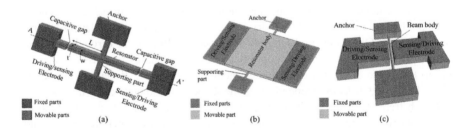

FIGURE 2.1   Resonator structures: (a) two-arm type resonator structure, (b) bar-type resonator structure, (c) beam-type resonator structure.

TABLE 2.1
Resonator Structure Parameters

Resonator Parameter	Length	Width	Thickness	Capacitive Gap
Symbol	L	W	t	g

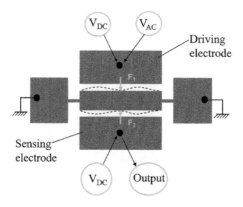

FIGURE 2.2 Electrostatically operated bar-type resonator.

deviance electrostatic force ΔF is the unbalance between F_1 and F_2, leading to the mechanical vibration, as shown in Equation 2.3.

$$F_1 = \frac{1}{2} \frac{C_f}{g-x} \left(V_{DC} + V_{AC}\right)^2, \tag{2.1}$$

$$F_2 = \frac{1}{2} \frac{C_f}{g-x} \left(V_{DC}\right)^2, \tag{2.2}$$

$$\Delta F = F_1 - F_2 = \frac{1}{2} \frac{C_f}{g-x} \left(2 V_{DC} V_{AC} + V_{AC}\right)^2, \tag{2.3}$$

$$C_f = \varepsilon_r \varepsilon_0 \frac{A_{el}}{g-x}, \tag{2.4}$$

where C_f is the feed-through capacitance (working capacitance), which varies along the resonant body due to its displacement x. g is the distance between two plates called the capacitive gap. A_{el} is the area of the electrode plate, ε_r is the dielectric constant of the material between the plates (for an air environment), and ε_0 is the electric constant ($\varepsilon_0 \approx 8.854 \times 10^{-12}\,\text{Fm}^{-1}$).

Assuming $V_{AC} \ll V_{DC}$ the effective electrostatic force $F_{electrostatic}$, namely ΔF, can be simplified as in Equation 2.5:

$$F_{electrostatic} = \Delta F = \frac{C_f}{g} V_{DC} V_{AC} = \eta V_{AC}, \tag{2.5}$$

$$\eta = V_{DC} \frac{\partial C_f}{\partial g} = -V_{DC} \frac{C_f}{g}, \tag{2.6}$$

where η is a transduction factor in the capacitive resonators, which represents the energy coupling efficient. From Equation 2.5, the larger transduction factor η in the resonator is the more electrical energy converted into the mechanical domain, and consequently the bigger the force difference can be gained for the vibration in the capacitive resonators.

The resonant frequency f_0 of the fundamental extensional vibration is given by

$$f_0 = \frac{1}{2\pi} \sqrt{\frac{k_{\text{eff}}}{m_{\text{eff}}}}, \tag{2.7}$$

where

k_{eff} is an effective spring constant, and
m_{eff} is an effective mass.

2.2   EQUIVALENT CIRCUIT MODEL

Electromechanical resonators can be inferred from a classic lumped mass–spring–dashpot by a mass, a damping coefficient, and a stiffness [9, 10]. The oscillation is driven with particular frequencies by an alternating force F and can be described with the mechanical motional equation:

$$m\ddot{x} + \gamma\dot{x} + kx = F. \tag{2.8}$$

The resonators have distributed mass and spring. To exactly model the electromechanical resonator in the mechanical domain, an effective damping coefficient, mass, and spring stiffness of the resonator body in introduced in Equation 2.9, where x is the displacement of resonant body toward the electrode. γ_{eff} and m_{eff} are fixed values for specific resonator design:

$$m_{\text{eff}}\ddot{x} + \gamma_{\text{eff}}\dot{x} + k_{\text{eff}}x = F. \tag{2.9}$$

The total current in the system can be calculated as in Equation 2.10:

$$I = \frac{\partial CV}{\partial t} = V\frac{\partial C}{\partial t} + C\frac{\partial V}{\partial t} \simeq V_{\text{DC}}\frac{\partial C_{\text{f}}}{\partial t} + C_0\frac{\partial V_{\text{AC}}}{\partial t}, \tag{2.10}$$

with referring to Equation 2.6,

$$V_{\text{DC}}\frac{\partial C_{\text{f}}}{\partial t} = V_{\text{DC}}\frac{\partial C_{\text{f}}}{\partial x}\frac{\partial x}{\partial t} = \eta\frac{\partial x}{\partial t}. \tag{2.11}$$

I can be reformed approximately as in Equation 2.12, which is made up of two parts, as in Equation 2.13:

$$I \simeq \eta\frac{\partial x}{\partial t} + C_0\frac{\partial V_{\text{AC}}}{\partial t} = \eta\frac{\partial x}{\partial t} + j\omega C_0 V_{\text{AC}}, \tag{2.12}$$

$$I = I_{\text{m}} + I_{\text{AC}} \tag{2.13}$$

The total current consists of motional current part $I_m = \eta \dot{x}$ and normal current $I_{AC} = j\omega C_0 V_{AC}$. The normal current derives from the AC voltage over the gap capacitances, which can be eliminated through cancellation of parallel capacitance aimed for clear output signals. The motional current as the desired output signal derives from the mechanical vibration through the change of the gap size. Equation 2.14 shows the relationship between I_m and mechanical transducer velocity v, where the displacement x is assumed to be small compared to the gap size g:

$$I_m = \eta \frac{\partial x}{\partial t} \simeq \eta \dot{x} = \eta v. \tag{2.14}$$

Substituting Equation 2.14 into the mechanical motion (Equation 2.8) yields

$$\frac{m}{\eta} \frac{\partial I_m}{\partial t} + \frac{\gamma}{\eta} I_m + \frac{k}{\eta} \int I_m dt = F(t). \tag{2.15}$$

Substituting Equation 2.5 into Equation 2.15, gives

$$\frac{m}{\eta^2} \frac{\partial I_m}{\partial t} + \frac{\gamma}{\eta^2} I_m + \frac{k}{\eta^2} \int I_m dt = V_{AC}. \tag{2.16}$$

By reforming Equation 2.5 and 2.14, the motional current (signal current) represents the mechanical resonating velocity and the excitation AC voltage represents the actuation force as in Equation 2.17:

$$\eta = \frac{F_{\text{electrostatic}}}{V_{AC}} = \frac{I_m}{v}. \tag{2.17}$$

The direct electromechanical analogies for lumped translational system are converted in Table 2.2. The resonators based on capacitive transduction can be modeled in the electric domain with equivalent electrical parameters, namely, a motional resistance, an inductance, and a capacitance C connected in series. These motional parameters are not real electrical parameters, but electrical equivalent parameters.

TABLE 2.2

Direct Electromechanical Analogies for Lumped Translational Systems

Mechanical Quantity	Electrical Quantity
Force, F	Voltage, V
Velocity, v	Current, I
Displacement, x	Charge, q
Mass, m	Inductance, L
Compliance, $1/k$	Capacitance, C
Viscous resistance, γ	Resistance, R

The motivation to build the equivalent circuit model is to simplify the analytical work with all effects of physics, mechanics, and circuits in the same domain.

According to Table 2.2, the mechanical motion equation in Equation 2.8 is transformed in Equation 2.18:

$$L_m \frac{\partial I_m}{\partial t} + R_m I_m + \frac{1}{C_m} \int I_m dt = V_{AC}. \tag{2.18}$$

Figure 2.3 illustrates the equivalent circuit model of the capacitive resonators. It consists of the motional resistance R_m, motional inductance L_m, motional capacitance C_m, and feed-through capacitance C_f [3, 7]. The motional resistance R_m represents mechanical losses of vibration, the motional inductance L_m reveals mechanical inertia, and the motional capacitance C_m corresponds to mechanical stiffness.

Comparing Equation 2.16 with Equation 2.18, L_m, C_m, and R_m are presented in Equations 2.19, 2.20, and 2.21, respectively:

$$L_m = \frac{m_{eff}}{\eta^2}, \tag{2.19}$$

$$C_m = \frac{\eta^2}{k_{eff}}, \tag{2.20}$$

$$R_m = \frac{\sqrt{k_{eff} m_{eff}}}{Q \eta^2} = \frac{\sqrt{k_{eff} m_{eff}}}{Q V_{DC}^2 \varepsilon_0^2 L^2 t^2} g^4, \tag{2.21}$$

where
   Q is the quality factor of the resonator, and
   L and t are the length and thickness of resonator, respectively.

As shown in Equations 2.19, 2.20, and 2.21, the circuit components are relative to the effective spring k_{eff}, the effective mass m_{eff}, the Q factor, the electromechanical

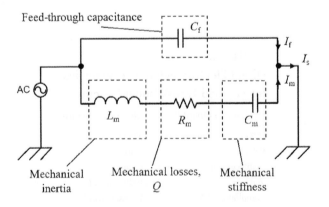

FIGURE 2.3   Equivalent circuit model.

transduction factor η, the area of the electrode plate, and the capacitive gap g. The motional resistance is considered as the most important parameters because it decides the effective performance of the silicon resonator. The large motional resistance results in high insertion loss and large phase noise that make capacitive resonators unable to meet oscillation conditions or oscillators with no stable time reference. The motional resistance is affected by many factors such as the polarization voltage V_{DC}, the overlap area of capacitance between resonant body and electrodes A_{el}, the Q factor, or the capacitive gap width g, but it strongly depends on the capacitive gap width with the fourth-order function (Equation 2.21). The lower value of the motional resistance can be obtained by reducing the capacitive gap width.

The mechanical resonance of the capacitive silicon resonator is represented by the series components called motional inductance L_m, motional capacitance C_m, and motional resistance R_m, and these are the equivalent of mass, spring, and damper, respectively. Due to the gap between the resonant body and electrodes, the capacitive silicon resonator exists as a feed-through capacitance that is associated with the equivalent circuit model as the aforementioned circuit (Figure 2.3). Therefore, there are two resonant frequencies that can be obtained by this equivalent circuit model: one is series resonant frequency f_s and the other is parallel resonant frequency f_P.

Equations for the resonant frequency f_s and the resonant frequency f_P are, respectively,

$$f_s = \frac{1}{2\pi\sqrt{L_m C_m}},$$ (2.22)

$$f_P = \frac{1}{2\pi}\sqrt{\frac{C_m + C_f}{L_m C_m C_f}}.$$ (2.23)

The equivalent circuit model (Figure 2.4a) of a capacitive silicon resonator is simulated by electronic circuit simulation using PROTEUS software (Labcenter Electronics Ltd.) and the result is shown in Figure 2.4b. Two resonant peaks have been observed including mechanical resonance and antiresonance. The motional components, including motional resistance, motional inductance, and motional capacitance, present mechanical properties of silicon resonator while feed-through is determined by the noise floor of sweep measurement and some additional capacitance between the input and output terminals. The effect of C_f is to introduce an antiresonance peak (parallel resonant frequency). The feed-through capacitance can drag down the measured response peak, and change the measured resonant frequency or even obscure the resonator response altogether. Thus, for high performance of capacitive silicon resonators, the feed-through gap (capacitive gap) should be as small as possible.

The accurate measurement of the motional component has been achieved by performing the feed-through capacitance cancellation. Figure 2.5 shows the frequency response of the capacitive silicon resonator after feed-through capacitance cancellation. The cancellation process is presented in detail in Chapter 5.

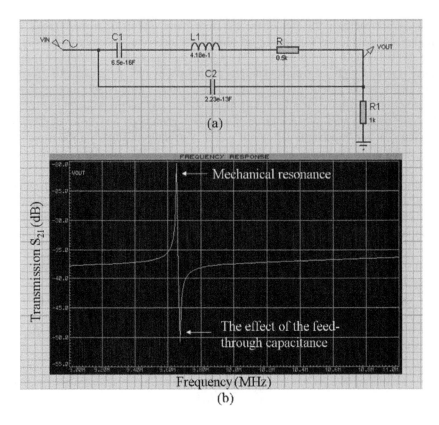

FIGURE 2.4   (a) Equivalent circuit model and (b) simulation result of capacitive silicon resonator.

2.3   FINITE ELEMENT MODEL

A finite element model (FEM) is built in COMSOL to predict vibration shapes and the operation frequency of the capacitive silicon resonators. To estimate the various physical domains of the problems, an electromechanics solver has been employed, which accounts for both the electrostatic and mechanical domains. By solving the stationary problem of the design, the displacement shapes can be found. By adding the eigenfrequency study to the stationary one, the fundamental frequency or high-order mode of the devices can be calculated.

The two-arm-type resonator structure is described in Figure 2.1a. The simulation of the mechanical resonance of the silicon resonator structure is performed by FEM, as mentioned earlier. The vibration shape at the fundamental frequency is the longitudinal extensional shape. The deformation in terms of changing the length of the resonant body is unity and longitudinal, which results in vibration in the length extensional mode. The deformation of the two-arm-type resonator is shown in Figure 2.6a. The bar-type resonator structure was shown in Figure 2.1b, and its vibration shape is shown in Figure 2.6b. At the fundamental frequency, the bar-type

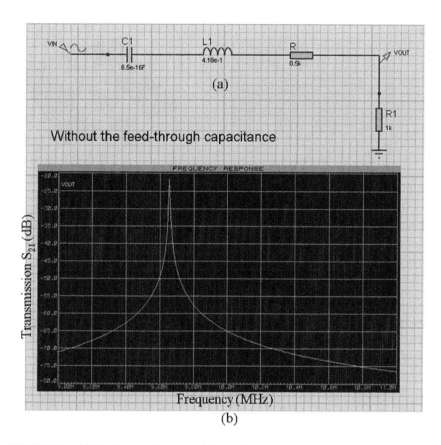

FIGURE 2.5 (a) Equivalent circuit model and (b) simulation result of capacitive silicon resonator without the feed-through capacitance.

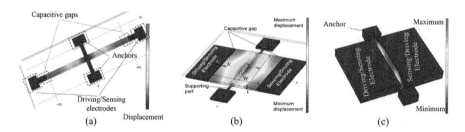

FIGURE 2.6 FEM simulation of (a) two-arm type capacitive silicon resonator, (b) bar-type capacitive silicon resonator, and (c) beam-type capacitive silicon resonator.

structure vibrates with the width extensional mode. Figure 2.6c shows the flexural vibration mode for the beam-type resonator structure.

The resonant frequency of capacitive silicon resonators mainly depends on their resonant body parameters. Resonant frequency can be determined by the width dimension for bar-type resonators, the length dimension for two-arm-type resonators,

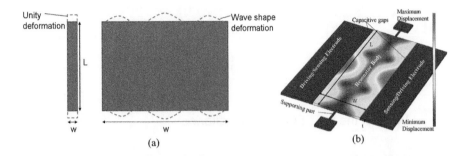

FIGURE 2.7   (a) Shape of the deformation, (b) Long-bar-type resonator structure.

and the beam dimension for beam-type resonators; therefore, resonators with a wide range of different frequencies can be fabricated simultaneously on the same silicon wafer. This is an advantage compared to piezoelectric resonators.

In bar-type resonators, the deformation shape depends on the aspect ratio (AR = Width/Length), as shown in Figure 2.7a. If the width w is less than half of the length L, the deformation of structure becomes unity. The unity displacement is observed at the end of the two arms for the two-arm-type resonator in Figure 2.6a. When the width increases, the two-arm-type resonator behaves as the bar-type resonator. A long-bar-type structure with wave-shaped deformation is shown in Figure 2.7b.

2.4   KEY PARAMETERS OF CAPACITIVE SILICON RESONATORS

2.4.1   RESONANT FREQUENCY

The resonant frequency of the resonators is typically determined by the physical dimensions of the resonator structure, its shape, the vibration mode, and the mechanical properties of the material. In the capacitive resonators in this book, the resonator structures are fabricated by silicon material and their resonant frequency is driven by the resonant body dimension. Resonant frequencies of the capacitive silicon resonators described in this work are located in the frequency range of 10 MHz to 100 MHz toward the application of timing devices.

2.4.2   QUALITY FACTOR

The Q factor can be derived from measured resonance peak, where the Q factor is defined as the ratio of the resonance frequency and the peak width at half maximum. The sharper the peak, the higher the Q factor. The Q factor can also be defined in terms of the ratio of the energy stored to the energy lost in a given cycle of oscillation. The Q factor is an important parameter for a resonator oscillator because it determines the motional resistance and phase noise. A higher Q factor helps to reduce motional resistance of the resonator and phase noise of the oscillator.

Energy loss comes from several loss mechanisms. Loss factors can either be external, that is, energy leaves the resonator and is dissipated outside of the resonator; or internal, that is, mechanical energy is directly dissipated inside or at the surface of the resonator. All of these factors can be combined to represent the total Q_{total} factor as a function of the Q factor due to each mechanism:

$$\frac{1}{Q_{total}} = \frac{1}{Q_{viscous}} + \frac{1}{Q_{anchor}} + \frac{1}{Q_{surface}} + \frac{1}{Q_{others}}, \qquad (2.24)$$

where

$Q_{viscous}$ is caused by viscous damping;

Q_{anchor} presents energy loss through the anchor;

$Q_{surface}$ is the surface effects and the Q factor is also effected by others elements; and

Q_{others} includes the material of the resonator structure, thermoelastic losses, etc.

2.4.3   MOTIONAL RESISTANCE

Motional resistance is one of the most important parameters for oscillator application. A large motional resistance R_m results in the high insertion loss and large phase noise that make silicon resonators unable to meet oscillation conditions or oscillations with no stable time reference. Moreover, a low R_m helps to obtain low-power oscillations. The oscillation condition is

$$R_m \leq \frac{1}{4\Pi f_0 C_f}. \qquad (2.25)$$

2.4.4   FEED-THROUGH CAPACITANCE

Capacitive silicon resonators consist of motional and feed-through components as shown in the equivalent circuit model, as in the aforementioned circuit (Figure 2.2). Feed-through capacitance is a physical capacitance formed by the gap between the resonant body and electrodes. The feed-through capacitance is in parallel with the motional component; consequently, there is a current path through the feed-through capacitance as shown in Figure 2.2. This current takes the current away from the motional component branch (mechanical resonant branch), which is equivalent to weakening the strength of the mechanical resonance. So, the impedance of the feed-through capacitance at resonance should be much larger than impedance of the resonance branch, such that the majority of the current goes through the mechanical resonant branch and not through the feed-through capacitance. Although the feed-through capacitance value is quite small, for high frequency applications, the electrical path through this capacitance is significant and could potentially obscure the resonant signal.

2.4.5   TUNING FREQUENCY

Tuning frequency is a property of silicon resonators due to changing the applied polarization voltage V_{DC}. It can be used to adjust the resonant frequency of resonant structure and potentially compensate temperature drifts of oscillation frequency. The tuning frequency for a silicon-bar-type resonator is shown in the following equation:

$$f = \frac{1}{2\pi}\sqrt{\frac{k_{\text{eff}}}{m_{\text{eff}}}} = \frac{1}{2\pi}\sqrt{\frac{k_{\text{m}}+k_{\text{e}}}{m_{\text{eff}}}} = \frac{1}{2\pi}\sqrt{\frac{k_{\text{m}} - \dfrac{V_{\text{DC}}^2 \varepsilon_0 L t}{2g^3}}{m_{\text{eff}}}}, \quad (2.26)$$

where k_{m} and k_{e} are mechanical and electrical spring constants, respectively.

The resonant peak shifts to lower frequency if the polarization voltage V_{DC} is increased due to the effect of electrical stiffness. The effect of the electrical stiffness caused by the polarization voltage on the resonant frequency of the resonator is clearly explained by Equation 2.26. In order to compensate for temperature drifts of oscillation frequency by an electrostatic tuning method, a wide range of tuning frequencies is required. Capacitive gap width reduction seems to be as one of the best methods to increase the tuning frequency range because the tuning frequency range is proportional to the third order of the capacitive gap width (Equation 2.26). Other methods are increasing the V_{DC} or the length or thickness of the resonant body.

REFERENCES

1. Toan, N.V., Shimazaki, T., Ono, T., Single and mechanically coupled capacitive silicon nanomechanical resonators, *Micro & Nano Letters*, **11**, 591–594, 2016.
2. Toan, N.V., Shimazaki, T., Inomata, N., Song, Y., Ono, T., Design and fabrication of capacitive silicon nanomechanical resonators with selective vibration of a high-order mode, *Micromachines*, **8**, 312, 2017.
3. Toan, N.V., Miyashita, H., Toda, M., Kawai, Y., Ono, T., Fabrication of an hermetically packaged silicon resonator on LTCC substrate, *Microsystem Technologies*, **19**, 1165–1175, 2013.
4. Toan, N.V., Toda, M., Kawai, Y., Ono, T., A long bar type silicon resonator with a high quality factor, *IEEJ Transactions on Sensors and Micromachines*, **134**, 26–31, 2014.
5. Toan, N.V., Toda, M., Kawai, Y., Ono, T., Capacitive silicon resonator with movable electrode structure for gap width reduction, *Journal of Micromechanics and Microengineering*, **24**, 025006, 2014.
6. Toan, N.V., Nha, N.V., Song, Y., Ono, T., Fabrication and evaluation of capacitive silicon resonators with piezoresistive heat engines, *Sensors and Actuators A: Physical*, **262**, 99–107, 2017.
7. van Beek, J.T.M., Puers, R., A review of MEMS oscillators for frequency reference and timing applications, *Journal of Micromechanics and Microengineering*, **22**, 013001, 2012.
8. Lee, J.E.-Y., Seshia, A.A., 5.4-MHz single-crystal silicon wine glass mode disk resonator with quality factor of 2 million, *Sensors and Actuators A: Physical*, **156**, 28–35, 2009.
9. Hartog, J.P.D., *Mechanical Vibrations*, Dover Publications, New York, NY, 1934.
10. Thomson, W.T., *Theory of Vibration with Applications*, 4th ed., Prentice-Hall, Englewood Cliffs, NJ, 1993.

3 Fabrication Techniques for Capacitive Silicon Resonators

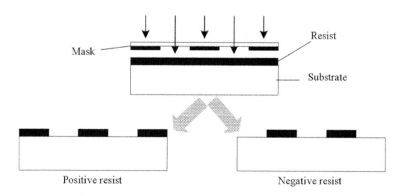

Micro/nanofabrication is the process of fabricating miniature structures such as microelectromechanical systems (MEMS) devices in micro- or nanoscale. To fabricate micro/nano devices, many processes must be conducted, one after the other, many times. These processes typically include depositing of a film, patterning the film with the desired micro/nano features, and removing portions of the film. This section gives an introduction of the micro/nanofabrication processes, experimental demonstrations, and discusses optimizing conditions that are used to fabricate the capacitive silicon resonators.

3.1 CLEANING PROCESS

The wafer cleaning process is one of the most importance processes in micro/nano fabrication. Contamination on a wafer's surface, which in general are referred to as particle contaminants or inorganic and organic residues, becomes a critical problem when device geometries become micro/nano scale. Small particles are hard to remove from silicon wafers due to strong electrostatic forces between particles and substrate. The yield of usable devices on the wafer is greatly influenced by the contaminants, which should be removed efficiently during the wafer cleaning process.

Generally, wafers are cleaned with weak acids to remove particle contaminants. The small particles can also be efficiently removed by using ultrasonic cleaning or fluid jet spray cleaning [1]. Piranha solution, which is a mixture of sulfuric acid (H_2SO_4) and hydrogen peroxide (H_2O_2), is widely used to clean metallic and organic residues off substrates. Additionally, another cleaning method, a process known as RCA cleaning, is based on hot alkaline and acidic hydrogen peroxide solutions. It was developed by Werner Kern in 1965 at Radio Corporation of America (RCA) [2–4]. Therefore, these procedures are also named as RCA cleaning. The general cleaning procedures performed in this work are as follows.

Piranha cleaning is a mixture made up of sulfuric acid (H_2SO_4) and hydrogen peroxide (H_2O_2). High concentrate sulfuric acid causes organic dehydration and carbonization; while hydrogen peroxide oxidizes carbonized product into carbon monoxide (CO) or carbon dioxide (CO_2) in gas phase. The principle of piranha cleaning is based on combining both reactions to remove carbon contaminants. Therefore, piranha solution is often used to clean photoresist and other organic residues, which is very hard to remove from wafers or chips. Piranha solution attacks most plastics violently, whereas polytetrafluoroethylene (PTFE) is safe for plastics. Piranha cleaning process should be done in containers or beakers made of glass using PTFE or stainless steel tools.

RCA-1 is a procedure for removing organic residues, some of the metal, and particles from substrate. Wafers are immersed in a mixture of ammonium hydroxide (NH_4OH) solvent, 30% H_2O_2 liquid, and deionized (DI) water with a volume ratio of 1:1:6 at 75°C for 30 min. Ions on the wafer are first oxidized by H_2O_2 creating a hydrophilic oxide layer in a few nanometers. The alkali (NH_4OH) solvent can get through the oxide layer and reacts (etches) slowly to oxide the silicon beneath, generating a soluble complex that can be rushed away with rinsing water. The oxidation and complex reactions happen in a circle, realizing the cleaning of wafers with slightly etching on the surface. Therefore, the surface roughness of the substrates degrades when using the RCA-1 process for a length of time.

RCA-2 is a procedure for removing metal ionic contaminants and alkaline from the substrate. Chips are immersed in a mixture of hydrogen chloride (HCl) acid, 30% H_2O_2 liquid, and DI water with a volume ratio of 1:1:6 at 75°C for 30 min. The solution can react with the metal before hydrogen (H) in the metal activities sequence, and form salt that can be dissolved in rinsing water.

Diluted HF is used to remove native oxide on a wafer's surface. In addition, particles are also removed by etching an underlying SiO2 layer to lift the particle away for the surface.

A summary of wafer cleaning processes is shown in Table 3.1.

3.2 DEPOSITION TECHNIQUES

3.2.1 SILICON DIOXIDE

The ability to form a chemically stable SiO_2 layer at the surface of silicon is one of the main reasons silicon is the most widely used semiconductor material. SiO_2 material possess a high quality electrically insulating layer, serving as dielectric in numerous micro/nano devices. In addition, SiO_2 material are often used as masks to etch silicon in the dry etching process. In order to form SiO_2 on a silicon surface, several methods have been investigated, including thermal oxidation and plasma-enhanced chemical vapor deposition.

TABLE 3.1
General Cleaning Procedures

Cleaning Processes	Chemicals	Conditions (Reaction Temperature; Process Time)	Removal
Piranha cleaning	96% H_2SO_4 acid, 30% H_2O_2 1:2 (volume ratio)	110°C; 30 min	Organics
RCA-1	NH_4OH liquid, 30% H_2O_2 , H_2O 1:1:6 (volume ratio)	75°C; 20 min	Organic residues, some metal, particles
RCA-2	HCL acid, 30% H_2O_2, H_2O 1:1:6 (volume ratio)	75°C; 20 min	Heavy metal ionic contaminants, alkaline
Diluted HF	HF (49 %) acid, H_2O 1:50 (weight ratio)	27°C; 20 sec	Native oxide

3.2.1.1 Thermal Oxidation Method

An SiO_2 film can be grown on the surface of silicon under high temperatures. The growth of the film is driven by the diffusion of oxygen into the silicon substrate. The basic thermal oxidation apparatus is shown in Figure 3.1. This apparatus consists of heater coils, a cylindrical fused-quartz tube containing the silicon wafers held vertically in a slotted-quartz boat, and a source of either pure oxygen or pure water vapor. The required oxygen can be supplied by flowing O_2. The temperature is raised to 900°C–1200°C to speed up the process.

There are two kinds of thermal oxidation processes. One is called dry thermal oxidation, which results in formation of SiO_2:

$$Si + O_2 \rightarrow SiO_2 \qquad (3.1)$$

Another is named wet thermal oxidation, where the required oxygen is supplied by water vapor instead of oxygen (Equation 3.2). The wet thermal oxidation process can be easily achieved with thick SiO_2 of around 1000 nm, whereas dry thermal takes a long time even around 500 nm.

$$Si + H_2O \rightarrow SiO_2 + H_2 \qquad (3.2)$$

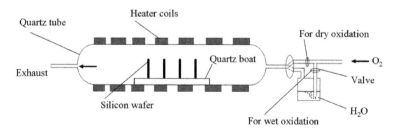

FIGURE 3.1 Schematic drawing of an oxidation system.

From Equations 3.1 and 3.2, the SiO_2 layer is formed by the diffusion of oxygen into the silicon substrate, which means that silicon is consumed by the film growth. Approximately 44% of the silicon surface is consumed during the oxidation process as shown in Equation 3.3 [5].

$$\frac{\text{Thickness of Silicon}}{\text{Thickness of } SiO_2} = \frac{\dfrac{\text{Molecular weight of Silicon}}{\text{Density of Silicon}}}{\dfrac{\text{Molecular weight of } SiO_2}{\text{Density of } SiO_2}}$$

$$= \frac{\dfrac{28.9 \text{ g / mol}}{2.33 \text{ g/cm}^3}}{\dfrac{60.08 \text{ g/mol}}{2.21 \text{ g/cm}^3}} = 0.44 = 44\%$$

(3.3)

3.2.1.2   Plasma-Enhanced Chemical Vapor Deposition

In general, a thick SiO_2 layer cannot be formed by thermal oxidation method owing to the problem of the diffusion of oxygen. Also, thick SiO_2 film exhibits a high stress film that causes buckling of the fabricated devices. As an alternative to thick SiO_2 film, plasma-enhanced chemical vapor deposition (CVD) using TEOS (tetraethoxysilane $Si(OC_2H_5)_4$) is a common selection. The TEOS molecule consists of a four-ethyl group (-CH_2CH_3) combined with an orthosilicate ion (SiO_4^{4-}). TEOS is a precursor of SiO_2 at high temperatures in the range of 100°C to 400°C. The formation of SiO_2 can be described as follows:

$$Si(OC_2H_5)_4 \rightarrow SiO_2 + 2(C_2H_5)_2 O$$

(3.4)

where diethyl ether $(C_2H_5)_2O$ is the evaporated co-product of the reaction.

In TEOS components, the Si atom is already oxidized, therefore, the changing from TEOS to SiO_2 is a process of a molecular rearrangement instead of an oxidation reaction. The TEOS CVD process can be conducted in an inert atmosphere or in ambience with an addition of O_2, which is used to increase the SiO_2 deposition rate. The reaction happens under a pressure of 0.6 Torr. The temperature of the process can be at 80°C (low-temperature recipe), 150°C (medium-temperature recipe), or 350°C (high-temperature recipe). The power of plasma is 70 W. The operating temperature mainly effects the stress of the deposited SiO_2 film. The low temperature results in low stress, but the quality of the SiO_2 is not good (e.g., there are many pinholes).

3.2.2   METAL DEPOSITION

3.2.2.1   Electron Beam Evaporation

The experimental setup for electron beam evaporation is shown in Figure 3.2. The electron beam evaporation setup is comprised of a hot cathode (filament) for

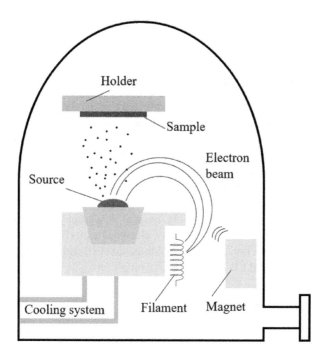

FIGURE 3.2 Schematic view of the electron beam evaporation setup.

launching electrons, an electron-accelerating electrode, and a coating source material as an anode. In high vacuum, the source materials are heated and evaporated by the hot electrons emitted for electron gun filament. With the accelerating electrode, the energy of the electron beam achieved on the source material can be highly concentrated, resulting in high local temperature of the source material, which evaporates with ease. By adjusting the power of the electron beam, it is possible to easily control the evaporation rate of the coating material, in particular, to favor the high melting point of pure metal and compounds. The thin film deposited by evaporation typically has tensile stress. Due to poor step coverage, this technique is usually used for lift-off processes.

3.2.2.2 Sputtering

The sputtering process utilizes plasma, which is generated at a low gas pressure (on the order of a few millitorrs) flowing across a closely spaced parallel electrode pair. High direct current (DC) voltage and a radio frequency (RF) generator are commonly used. The ion flow in the plasma is accelerated through the electric field, which is needed for bombardment of the target. Aimed for a better bombardment effect, a flow of relatively heavy ions with high kinetic energy is used to knock off atoms on the surface of the target material. The free atoms settle on the surface of the substrate with a certain velocity and gradually forms a film of the desired deposition layer. In general, Ar gas–induced plasma is chosen in the sputtering process because of its high atomic mass on one hand. On the other hand, Ar is an inert gas, which does not react with the target or substrate. Moreover, Ar has a much lower price than other high atomic insert gases such as krypton (Kr) or xenon (Xe).

3.3   LITHOGRAPHY

Lithography is the process of transferring a mask pattern onto a substrate, which is the one of the most essential processes in microelectronics and MEMS fabrications. The lithography process generally consists of the following steps: prebaking of the resist for resist hardening; resist exposure in a lithography tool; development of photoresist; and postbaking of the resist to fully harden it. Different lithography processes are named according to the exposure mechanism happening inside the lithography tools, thereby special resists and different working procedures are developed to meet the exposure requirements. Several lithography technologies are currently available including photolithography, electron beam (EB) lithography, optical ultraviolet (UV) laser writer, and focused ion beam (FIB) lithography. In this work, the photolithography and EB lithography are employed.

3.3.1   PHOTOLITHOGRAPHY

Photolithography uses light (photons) to transfer a pattern from a mask onto a light-sensitive resist on the substrate. Light from the UV source (a mercury lamp) is reflected by mirrors and illuminated on the mask. The mask located between the UV source and sample has desired patterns, so that UV light only exposes a specific area on the resist. During the exposure process, the polymer bonds in the resist are chemically damaged by the dissipated energy from the lithography medium. The photolithography procedure is described as follows:

The wafer cleaning is standard, including piranha cleaning and RCA cleaning.
Baking in 145°C oven for 30 min to remove the adsorbed water from the wafer surface.
The photoresist coating is added by a spin coater.
Prebaking removes solvents and stress, and improves the adhesion of the photoresist on the wafer.
Exposure. After alignment between the wafer and mask, the photoresist is exposed through the pattern on the mask with a high-intensity UV light within the determined time calculated from the exposure dose of the photoresist to transfer the mask images onto the photoresist.
Development is a process to transform the latent resist image formed during exposure into a resist pattern on the surface of the substrate [3]. If the exposed resist dissolves in the developer due to chain scission, it is called a positive resist. In contrast, a negative resist becomes more chemically stable through bonds cross-linking after being exposed. Figure 3.3 illustrates the difference between positive and negative resist after exposure and development.
Postbaking improves the hardness and increases the adhesion of the resist to the wafer surface.
Oxygen plasma etching removes unwanted resist left behind after development.

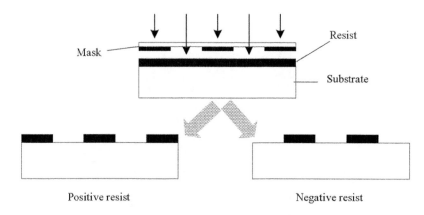

Positive resist   Negative resist

FIGURE 3.3   Photolithography process with positive and negative photoresists.

3.3.2 ELECTRON BEAM LITHOGRAPHY

Electron beam lithography (EBL) is one of the most useful tools for patterning small devices with nanometer resolution. EBL refers to a lithographic process that uses a focused beam of electrons to form fine patterns on the substrate, in contrast with optical lithography, which uses light for the same purpose. EBL offers higher patterning resolution than optical lithography due to the shorter wavelength possessed by the 10 KeV–100 KeV electrons that it employs. The resolution of optical lithography is limited by diffraction, but this is not a problem for EBL because of the short wavelengths (0.2–0.5 angstroms). EBL has the benefit of extremely high diffraction-limited resolution and has been used for transferring patterns of nanometer size. However, the resolution of EBL may be constrained by other factors, such as electron scattering in the resist and aberrations in its electron optics. Just like optical photolithography, EBL also uses positive and negative resists. In the positive resist, the area exposed under the electron beam will be removed after development and the unexposed area will remain. On the other hand, in a negative resist, the area exposed under the electron beam will remain after development. An extra postbake step is also required in EBL. The minimum feature that can be resolved by the EBL depends on several factors, including type of resist, resist thickness, exposure dosage, the beam current level, proximity correction, and development process. In this research, the JEOL EB 5000LSS is used to obtain the narrow gap with a high resolution of positive photoresist (ZEP 520A). The EBL conditions are shown in Table 3.2 and the EBL results are in Figure 3.4. The effects of dose times and gap widths are shown in Figure 3.5.

TABLE 3.2
EB Lithography Conditions

EB Resist	Thickness of EB Resist	Beam Current	High Voltage	Dose Time
ZEP 520A	Around 400 nm	100 pA	50 kV	85 µC/Cm^2

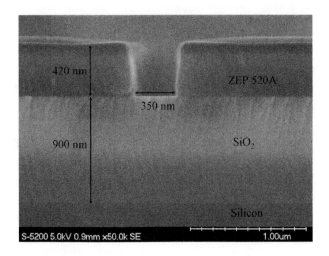

FIGURE 3.4   EB lithography results.

3.4   ETCHING TECHNIQUES

3.4.1   Silicon Dioxide and Glass

Silicon dioxide (SiO_2) and glass materials have been widely used for micro/nano systems because of transparency, mechanical robustness, and dielectric properties; however, they are difficult to machine precisely in micro/nano scales. Several etching techniques have been investigated in this work, as follows.

3.4.1.1   Sandblasting

Sandblasting is a technique in which a particle jet is directed toward a target (sample) for material removal by mechanical erosion via the impingement of high velocity abrasive particles. The sandblast process can be used for etching various materials such as glass [6], ceramics [7] (e.g., low temperature co-fired ceramics [LTCC]), and silicon.

Figure 3.6 illustrates the sandblasting setup, which commonly consists of a nozzle, a micropowder, and a movable stage. The particles are accelerated toward the sample with high-pressure airflow through the nozzles of sandblaster. The etching rate of the sample is controlled by the jet velocity of powder commonly at 80–290 m/s and the movement velocity of the stage (X velocity and Y velocity). In this work, the sandblasting process is used for the patterning of the Tempax glass, using Al_2O_3 powder with a granule size of 14 μm.

Figure 3.7 illustrates the experimental process of sandblasting on the glass wafer. A 300-μm thick glass substrate (Figure 3.7a) is employed for this process. A dry film resist (MS 7050, Toray, Japan) with a thickness of 50 μm is pasted, and then photolithography is performed, as shown in Figure 3.7b. Next, glass wafers with dry film resist patterns are etched via sandblasting. Illustration images for the sandblasting results are shown in Figure 3.8. A 2 cm × 2 cm LTCC and glass wafers are sandblasted, as shown in Figure 3.8a and b, respectively. The glass etching surfaces are very rough and etching profiles evolve into V-shapes.

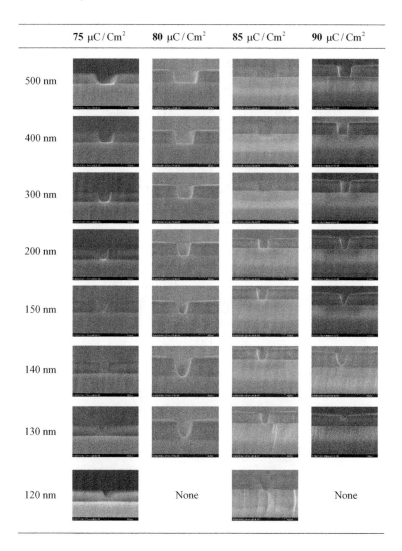

FIGURE 3.5 EB photolithography profile with different dose times and gap widths.

3.4.1.2 Wet Etching

The wet etching of glass has been investigated by many researchers [8, 9]. The advantages of this method include its simplicity, high etching rate, high mask selectivity, and low surface roughness. However, due to its isotropic etching behavior, the aspect ratio is limited. A buffered hydrofluoric acid (a diluted HF with ammonium fluoride [NH_4F]) solution is used for the etching of SiO_2 because of the low damage to the photoresist. Therefore, durability to the etching solution is improved. A higher etching rate can be achieved by increasing the concentration of the HF solution; however, the quality of the photoresist mask becomes poor. Thus, it may only be suitable for the etching of thin SiO_2 layers. To overcome this problem, Cr–Au is one selection for

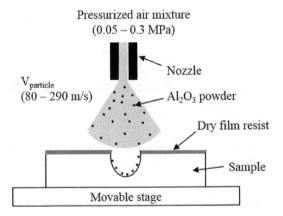

FIGURE 3.6   Sandblast setup.

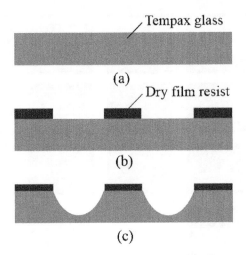

FIGURE 3.7   Fabrication process. (a) Tempax glass, (b) photolithography, (c) sandblasting.

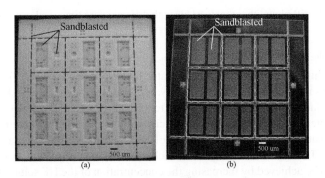

FIGURE 3.8   Illustration images for the sandblasting process. (a) Sandblasting result on LTCC wafer. (b) Sandblasting result on Tempax glass.

the masking material for wet etching due to the inert property of Au when in contact with HF. The etching of glass has more difficulties than that of SiO_2 due to its composition. Therefore, deep etching requires long etching times. In this experiment, the Tempax glass was etched using a diluted solution of 50% HF:DI = 2:1. Summarized wet etching conditions are shown in Table 3.3. The chemical reaction of the glass in the HF solution is as follows:

$$SiO_2 + 4HF = SiF_4 + 2H_2O \qquad (3.5)$$

The wet etching process starts from a 300-μm thick glass substrate (Figure 3.9a). A Cr layer 30 nm thick and an Au layer 300 nm thick are deposited on both sides of the Tempax glass wafer via sputtering (Figure 3.9b). Conventional photolithography, using a photoresist (OFPR 200 cp), is performed on the front side to make the mask pattern. The same photoresist is coated on the back side (Figure 3.9c). Then, Cr–Au layers are etched by the wet etchant (Figure 3.9d). The experimental result is shown in Figure 3.10a. Finally, the wafer is dipped in the etching solution of the diluted HF (Figure 3.9e). The glass structure with an etching depth of approximately 20 μm was achieved by the aforementioned solution and an etching time of 10 min. The side etching of the glass with a length of about 20 μm was observed as shown in Figure 3.10b.

3.4.1.3 Fast Atom Beam Technology

A fast atom beam (FAB) used for dry etching is capable of high anisotropic etching because it utilizes neutral etching species. It generates the energetic neutral particles for bombardment of etching various materials such as silicon, silicon dioxide SiO_2, and metal. The advantage of this technique it that it enables fabricating fine, precise structures for high anisotropy etching [10]. Figure 3.11 illustrates the principle and configuration of the FAB apparatus [11]. In this section, the etching results of SiO_2 using FAB technology are introduced. The FAB etching of SiO_2 has been carried out by using CHF_3 gas with high selectivity. The etching rate of SiO_2 depends on the gas species, gas pressure, wafer temperature, and discharge conditions.

In our experiment, the etching conditions are as following: a CHF_3 gas flow rate of 8 sccm is used for etching, the anode–cathode voltage is 2.5 kV, the discharge current is 25 mA, and the etching rate of SiO_2 is around 10 nm/min. The etching profile of FAB technology gives a V-shape profile with fine etching surfaces as shown in Figure 3.12. Better etching results with vertical sidewalls may be obtained by optimizing the etching conditions.

TABLE 3.3
Wet Etching Conditions

Etching Material	Mask Material	Etching Solution	Etching Rate	Side Etching
Tempax Glass	Photoresist on metal (Cr–Au)	HF:DI = 2:1	2 μm/min	2 μm/min

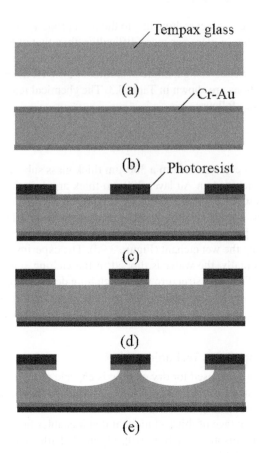

FIGURE 3.9   Wet etching process. (a) Tempax glass, (b) Cr–Au sputter, (c) photolithography, (d) Cr–Au wet etching, (e) glass wet etching.

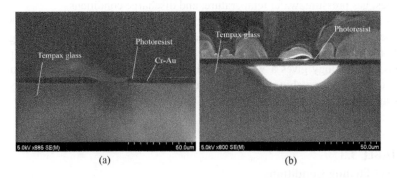

FIGURE 3.10   (a) Photolithography and metal etching; (b) glass etching result.

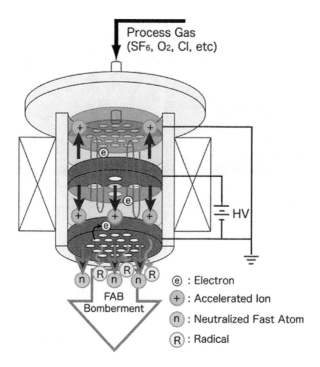

FIGURE 3.11 Schematic structure of FAB apparatus.

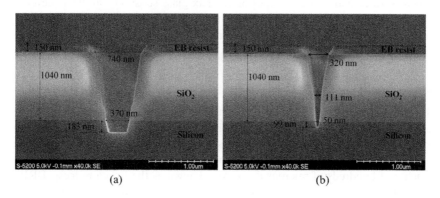

FIGURE 3.12 The etching profile of FAB technology: (a) large gap width, (b) narrow gap width.

3.4.1.4 Reactive Ion Etching

Reactive ion etching (RIE) using RF glow discharges and a reactive gas such as CHF_3, CF_4, or SF_6 is a dry etching technique for silicon, silicon dioxide, metals, and other materials. Gas molecules are broken down into many fragments and radicals in the plasma environment. These molecular fragments are ionized in the plasma and accelerated to the electrode surface within the discharge chamber. In RIE systems,

(a)   (b)

FIGURE 3.13   RIE results: (a) SiO_2, (b) Tempax glass.

the target substrate is electrically bias connected and attacked by the flux of ions or radicals from reactive gas. Although the high-density plasma help improves the efficiency, the etching becomes more isotropic as well.

A thick SiO_2 layer can be etched by the RIE method to form the etching mask of silicon. Good etching results with vertical sidewalls and narrow gaps are obtained by choosing the right etching conditions. In our experiment, the etching process of SiO_2 is carried out using an EB resist mask, a mixture gas of CHF_3 and Ar at a power of 120 W, and a chamber pressure of 5 Pa. The etching profile of RIE technology is shown in Figure 3.13a. In addition, a high-aspect-ratio structure together with a smooth surface and vertical shapes of Tempax glass substrate can be achieved via RIE, as shown in Figure 3.13b. This technique is a potential candidate for micro/nano scale glass micromachining. More information can be found in our previous work [12, 13].

3.4.2 SILICON

The patterning of silicon structures, which has been studied and developed for years, can be done by both wet (KOH and TMAH [tetramethylammonium hydroxide]) and dry (reactive ion etching) etching methods. Forming high-aspect-ratio silicon structures can be easily achieved by DRIE, which allows achievement of almost vertical trenches of hundreds of microns deep. Also known as the Bosch process, DRIE is time-multiplexed etching, which alternates between two modes, each lasting a few seconds. The Bosch process is explained as follows [14]:

Coating of the entire surface with "polymer." A Teflon-like substance is formed in the plasma from the C_4F_8 gas.

The C_4F_8 is turned off and SF_6 is let into the chamber. The SF_6 plasma etches the polymer in the bottom of the trench and once that is removed completely, the SF_6 etches the silicon in the bottom of the trench. The SF_6 does not etch the polymer on the walls because to etch polymer requires both radicals and ions.

The cycle is repeated until the structure has been etched.

The photoresist, SiO_2, or Si_3N_4 can be used as the mask to get silicon trenches with a high aspect ratio since the selectivity to silicon is high. Various materials with etching selectivity to silicon in the DRIE are shown in Table 3.4.

In this research, narrow gaps with small scallops are required for capacitive silicon resonators. To achieve this requirement, the Bosch process should be optimized. The short etching and passivation cycles (etching time 4 sec and passivation 3 sec) should be used for the Bosch DRIE process that can create small scallops of around 20 nm, as shown in Figure 3.14. It should also be noted that the selectivity between the material mask and silicon is drastically decreased by these Bosch DRIE conditions. Therefore, a hard mask (e.g., SiO_2) material should be selected to get higher selectivity in a comparison with soft mask (e.g., photoresist). To etch 5 μm of a silicon device layer with small scallops, the Bosch DRIE requires at least 300 nm layer of SiO_2 in our experiments.

TABLE 3.4
Various Materials with the Etching Selectivity to Silicon in DRIE

Materials	Selectivity to Silicon	Remarks
OFPR800-200cp	1:100	Positive photoresist
SiO_2	1:350	Thermal oxidation
SiO_2	1:250	TEOS CVD
Si_3N_4	1:150	—

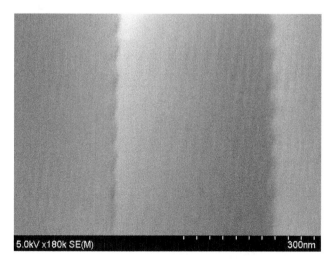

FIGURE 3.14 Bosch DRIE process result with short etching and passivation cycles.

3.5 ANODIC BONDING PROCESS

There are many different ways to bond silicon wafers with different materials. One of the popular ways to bond silicon wafers with other material is anodic bonding, which was discovered by Willis and Pomerantz [15]. The anodic bonding process has a variety of commercial applications including pressure sensors, photovoltaics, and microelectronic device packaging. It has been used to bond many different materials such as ceramics with metal, ceramics with semiconductors, ceramics with ceramics, and semiconductors with semiconductor pairs. In anodic bonding, the process relies on charge migration to produce bonded wafers. This usually works well for elements with high alkali metal content. Silicon wafer is one of the main materials used in the bonding process. Glass is another material that is high in alkali metal, therefore it is often used in the anodic bonding process to bond silicon to glass. Figure 3.15 is a schematic of the mechanism of anodic bonding. In this work, an anodic bonding process to bond silicon wafer and LTCC is performed with a temperature of approximately 400°C and an electric field of approximately 0.8 kV. The anodic bonding setup for LTCC and SOI is shown in Figure 3.16. The silicon and LTCC wafers are aligned and fixed on a sample holder, as shown in Figure 3.17.

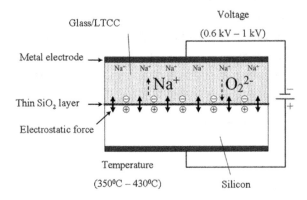

FIGURE 3.15   Schematic of mechanism of anodic bonding.

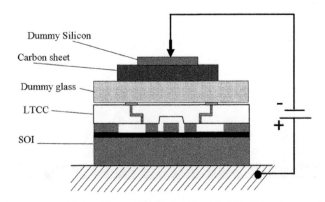

FIGURE 3.16   Anodic bonding setup in this research.

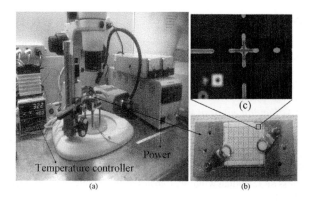

FIGURE 3.17 (a) Anodic bonding setup. (b) LTCC and SOI are aligned and fixed on a sample holder. (c) Close-up image of alignment area.

3.6 SUPERCRITICAL DRYING

Supercritical drying is a process to remove liquid in a precisely controlled way without severe deformation and collapse of structure. The primary cause of such damage is the effect of surface tension. The specimen is subject to considerable forces, which are present at the phase boundary as the liquid evaporates. The most common specimen medium, water, has a high surface tension to air, by comparison to that for acetone, which is considerably lower. However, after wet processing (such as release), a specimen dried by supercritical drying is unaffected by any surface tension, thus preventing stiction. The common transitional method is liquefied CO_2 due to its low critical temperature and the associated critical pressure. For this reason, the supercritical CO_2 has extremely low surface tension, and as a result does not pull structures down as it transitions from liquid to gas. The supercritical dryer machine is introduced in Figure 3.18.

FIGURE 3.18 Supercritical dryer machine.

REFERENCES

1. Hirano, H., Rasly, M., Kaushik, N., Esashi, M., Tanaka, S., Particle removal without causing damage to MEMS structure, *IEEJ Transactions on Sensors and Micromachines*, **133**, no. 5, 157–163, 2013.
2. Kern, W., *Handbook of Semiconductor Wafer Cleaning Technology*, Noyes Publications, Park Ridge, NJ, 1983.
3. Madou, M.J., *Fundamentals of Microfabrication: The Science of Miniaturization*, 2nd ed., CRC Press, Boca Raton, FL, 2002.
4. Petersen, K.E., Dynamic micromechanics on silicon: Techniques and devices, *IEEE Transactions on Electron Devices*, **25**, 1241–1250, 1978.
5. Toan, N.V., Sangu, S., Ono, T., Design and fabrication of large area freestanding compressive stress SiO2 optical window, *Journal of Micromechanics and Microengineering*, **26**, 075016, 2016.
6. Wensink, H., Berenschot, J.W., Hanse, H.V., Elwenspoel, M.C., High resolution powder blast micromachining. In: *Proceedings of the Micro Electro Mechanical Systems*, Miyazaki, Japan, 23–27 January, pp. 769–774, 2000.
7. Toan, N.V., Miyashita, H., Toda, M., Kawai, Y., Ono, T., Fabrication of an hermetically packaged silicon resonator on LTCC substrate, *Microsystem Technologies*, **19**, 1165–1175, 2013.
8. Grosse, A., Grewe, M., Fouckhardt, H., Deep wet etching of fused silica glass for hollow capillary optical leaky waveguides in microfluidic devices, *Journal of Micromechanics and Microengineering*, **11**, 257–262, 2001.
9. Iliescu, C., Chen, B., Miao, J., On the wet etching of Pyrex glass, *Sensors and Actuators A: Physical*, **143**, 154–161, 2008.
10. Ono, T., Orimoto, N., Lee, S., Simizu, T., Esashi, M., RF-plasma-assisted fast atom beam etching, *Japanese Journal of Applied Physics*, **39**, 6976–6979, 2000.
11. Shimokawa, F., Kuwano, H., Energy distribution and formation mechanism of fast atom in a fast atom beam, *Journal of Applied Physics*, **72**, 13–17, 1992.
12. Toan, N.V., Sangu, S., Ono, T., Fabrication of deep SiO2 and Tempax glass pillar structures by reactive ion etching for optical modulator, *Journal of Microelectromechanical Systems*, **25**, 668–674, 2016.
13. Toan, N.V., Toda, M., Ono, T., An investigation on etching techniques for glass micromachining, *Micromachining*, **7**, 51, 2016.
14. A deep silicon RIE primer: Bosch etching of deep structures in silicon, 2009, http://www.nanofab.ualberta.ca/wp-content/uploads/2009/03/primer_deepsiliconrie.pdf.
15. Wallis, G., Pomerantz, D.I., Field assisted glass-metal sealing, *Journal of Applied Physics*, **40**, no. 10, 3946–3949, 1969.

Part 2A

Performance Enhancement Methods for Capacitive Silicon Resonators

Fabrication Technologies

4 Hermetically Packaged Capacitive Silicon Resonators on LTCC Substrate

Chip 2x2 cm after packaging

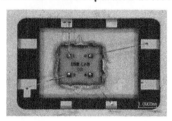

Font side of the separated device

Back side of the separated device

Mounted on 4 pin DIP and Au wire bonding

4.1 INTRODUCTION

The packaging process [1] for encapsulating and electrical interconnections is a critical technology for practical applications of microfabricated resonators. The packaging must not only avoid viscous damping, which can prevent the high-quality factor and high stability of the resonance, but it must also have the ability to protect the device from the external environment and to avoid issues with moisture and particles detrimental to long-term operation. The successful integration of the resonator with large-scale integration (LSI) can easily compensate the temperature drift of the resonator by using a temperature compensation circuit.

Encapsulation using a thin-film package was presented by Pourkamali and Ayazi [2]. The use of a high-temperature process for the deposition of the thin film and the release of a significant amount of hydrogen are the disadvantages of this method. A thin-film packaging process based on a metal–organic thin film [3] is available at low temperature (<110°C), but it also suffers from process incompatibilities. Although getters are typically employed for the vacuum packaging to absorb the trapped and desorbed gases during the sealing process, to become active getters need to be heated in a vacuum at temperatures of 300°C to 500°C. Packaging using a conventional borosilicate glass substrate with electrical feed-through has been demonstrated [4, 5]. Nevertheless, the quality of via holes with metal is poor. The packaging process using a low temperature co-fired ceramic (LTCC) substrate offers great potential to reduce cost and improve reliability. This substrate is a multilayer, glass ceramic substrate that contains the metal feed-through and can be anodically bonded to silicon. The sealed packaging of radio-frequency microelectromechanical systems (RF MEMS) produced by the anodic bonding process using LTCC has been reported [6–8].

Silicon resonators with hermetic packaging based on the utilization of the LTCC substrate were proposed for the integration of the resonator with LSI, as presented in our works [9–12]. The resonator structure is transferred onto the LTCC substrate using the anodic bonding between silicon and LTCC for electrical interconnections, and the resonator structure is packaged hermetically by the second anodic bonding of the silicon and Tempax glass. These works bring out a new approach to combine vacuum packaging technology with the silicon resonator on the LTCC substrate via the anodic bonding method. Figure 4.1a and b describes the titled and cross section views of the proposed capacitive silicon resonators, respectively.

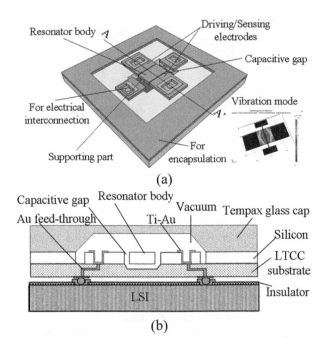

(a)

(b)

FIGURE 4.1 Proposed silicon resonant structure capable of the integration of LSI. (a) Bar-type resonator on the LTCC substrate with thin-film metal interconnections. (b) A–A' cross section of the device structure (after hermetically packaged).

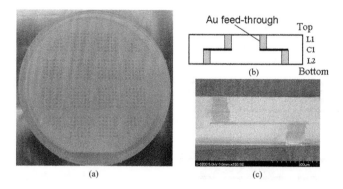

FIGURE 4.2 LTCC substrate. (a) 4-inch LTCC wafer. (b) Sketched cross section of feed-through vias. (c) SEM image of a cross section of the feed-through.

4.2 LTCC SUBSTRATE

LTCC substrate (Nikko Co.) is a special substrate that can be anodically bonded to silicon. It is composed of alumina, cordierite, and $Na_2O-Al_2O_3-B_2O_3-SiO_2$ glass. Figure 4.2a shows the 4-inch LTCC wafer. The sketched cross section of the feed-through vias and its finite element model (FEM) image are shown in Figure 4.2b and c, respectively.

The thermal expansion coefficient (CTE) of the LTCC is not only matched to that of silicon, but its special composition also allows it to be anodically bonded with silicon. The LTCC has higher reliability at high temperature in comparison with a conventional borosilicate glass substrate with metal vias because screen-printed metal vias in the LTCC substrate contain fillers relaxing the coefficient of thermal expansion mismatch between the vias and the substrate itself.

4.3 DEVICE FABRICATION

Silicon resonator structures were fabricated by following the fabrication process shown in Figure 4.3. The details of the fabrication process are as follows.

The fabrication process was started from a silicon on insulator (SOI) wafer (Figure 4.3a), which consists of a 7 µm-thick top silicon layer, 1 µm-thick oxide layer, and 300 µm-thick silicon handling layer. A 300 nm-thick SiO_2 was grown on the SOI by dry oxidation for the etching mask of silicon. On this SiO_2 layer, a resist pattern was formed by electron beam (EB) lithography, and then the SiO_2 layer is etched by using a fast atom beam (FAB) of SF_6 gas. The capacitive gaps were formed by using deep reactive ion etching (DRIE) based on the Bosch process. This process can create small scallops (~50 nm) using short etching and passivation cycles (etching time, 4 sec; passivation time, 3 sec). Figure 4.3b and b' presents the A–A' cross section and top view of the pattern after EB lithography, FAB, and DRIE processes, respectively. Next, the resonator structure was created by a photolithography process and DRIE process. Figure 4.3c shows the A–A' cross section of the pattern and there are some undesired parts, as shown in Figure 4.3c' and Figure 4.4a. The fabricated

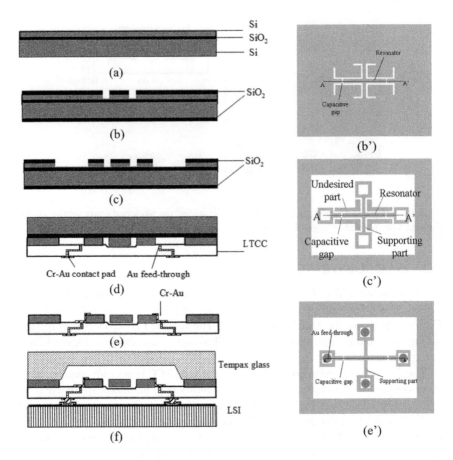

FIGURE 4.3 Fabrication process of a silicon resonator on the LTCC substrate with LSI integration. (a) SOI wafer – 7 μm/1 μm/300 μm. (b) After EB lithography, FAB, and DRIE process. (b') Top view of the pattern after EB lithography, FAB, and DRIE process. (c) Photolithography and DRIE process. (c') Top view of the pattern after photolithography and DRIE process. (d) First anodic bonding: silicon and LTCC. (e) Buried SiO_2 layer and electrical connections. (e') Top view of the pattern after removing of handling silicon layer and buried SiO_2 layer. (f) Second anodic bonding and LSI integration.

result of the two-arm-type resonator with a capacitive gap size of 500 nm is also shown in Figure 4.4a.

Before the anodic bonding process, Cr–Au contact pads with thicknesses of 20 nm and 300 nm, respectively, were formed by using the wet etching process with a photoresist mask on the back side of the LTCC substrate, and the dry film photoresist was patterned on the LTCC substrate as a mask for sandblasting. The LTCC substrate was sandblasted using Al_2O_3 to form approximately 100 μm-deep cavities at the underlining position of the resonator for making freestanding structures. The Cr–Au patterns on the back side of the LTCC are shown in Figure 4.4b, and the results of the sandblasting process of the LTCC substrate are presented in Figure 4.4b and c.

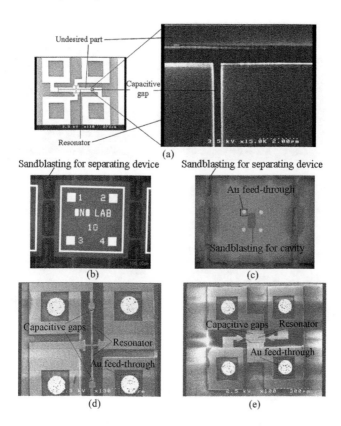

FIGURE 4.4 (a) SEM view of two-arm-type resonator with 500 nm gap width after EB lithography, photolithography, and DRIE process. (b) Back side of LTCC with Cr–Au layer for bonding with LSI. (c) Front side of LTCC with a cavity for freestanding structure. (d) SEM image of the two-arm-type resonator on LTCC substrate. (e) SEM image of the bar-type resonator on LTCC substrate.

The SOI wafer and the LTCC substrate are aligned and bonded together at 400°C with an applied voltage of 800 V for 5 min (Figure 4.3d). The handling silicon layer of the SOI wafer is removed by plasma etching using SF_6 gas. The undesired parts mentioned earlier will be taken out automatically after the buried SiO_2 layer is removed by the FAB of CHF_3 gas. The resonator structures shown in Figure 4.4d and e were transferred from the SOI wafer to the LTCC substrate.

Then, the Cr–Au films with thicknesses of 20 nm and 300 nm, respectively, were partly deposited by the sputtering process via a stencil mask for electrical connections between the silicon electrodes and the metal feed-through of the LTCC (Figure 4.3e). The second anodic bonding process was performed for hermetic sealing between the Tempax glass cap and silicon. Figure 4.5a shows the image of a completed resonator structure with LTCC integration for packaging of 36 2 × 2 cm² silicon wafers. After packaging, the devices can be separated by mechanical scribing. The separated device size is 2 × 2 × 0.6 mm³, as shown in Figure 4.5b and c,

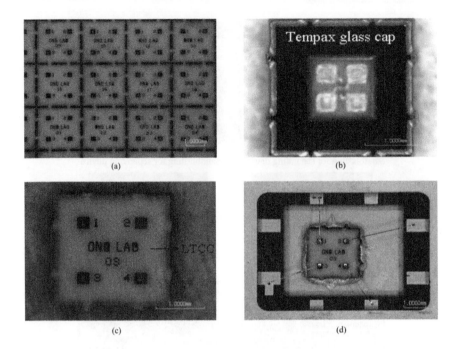

(a)   (b)

(c)   (d)

FIGURE 4.5 Micrographs of the device after packaging process. (a) Chip with a size of 2×2 cm² after packaging. (b) Front side of the separated device. (c) Back side of the separated device. (d) Device mounted on a 4-pin DIP and bonded with Au wire.

which is the front side and the back side of the separated device in the direction of Tempax glass and the LTCC substrate, respectively.

The device can be bonded with LSI for the oscillation and temperature compensation circuits using a solder bonding technique. The current packaged device mounted on a 4-pin dual in-line package (DIP) and Au wire bonding for characteristic measurement is shown in Figure 4.5d.

4.4 MEASUREMENT SETUP

Resonant performances were measured using a network analyzer (Anritsu MS4630B) in a range from 10 Hz to 300 MHz. The measurement setup for the characterization of the capacitive resonators is shown in Figure 4.6. A DC voltage was applied to the driving and sensing electrodes against the grounded resonator through a 100 kΩ resistor, which was decoupled from the radio frequency (RF) output of the network analyzer using a 100 nF capacitor.

The resonator structures were mounted on a 4-pin DIP using epoxy. The driving electrode, sensing electrode, and anchor pad were then wire-bonded to the gold pins of the DIP, which were soldered to the printed circuit board (PCB) with coaxial connectors, and they are connected to the network analyzer and the DC supply.

The resonator structure formed on the LTCC substrate with a packaging process can be directly mounted on the DIP using epoxy (Figure 4.7a), whereas that without

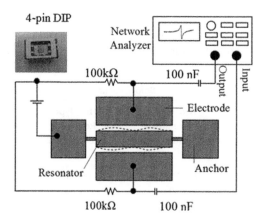

FIGURE 4.6   Measurement setup.

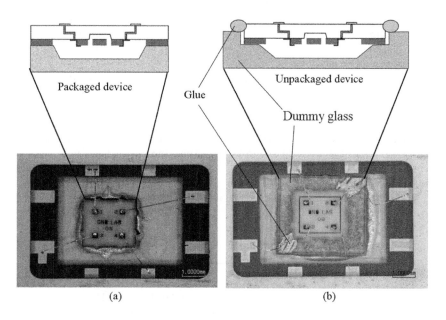

FIGURE 4.7   Device mounted on a 4-pin DIP and bonded with Au wire. (a) Packaged device. (b) Unpackaged device.

a packaging process needs the dummy glass to perform electrical connections for the characterization of the devices and also to avoid particles, as shown in Figure 4.7b.

4.5   MEASUREMENT RESULTS

The silicon resonators formed on the LTCC substrate without a packaging process are placed inside a vacuum chamber with coaxial feed-through. The resonant characteristics of a two-arm-type resonator (resonator 1) and bar-type resonator (resonator 2)

TABLE 4.1

Summarized Resonant Characteristics of Some Types of Silicon Resonators

	Before Packaging		After Packaging
	Two-Arm-Type Resonator (Resonator 1)	Bar-Type Resonator (Resonator 2)	Two-Arm Type Resonator (Resonator 3)
Resonator Parameters			
Length	$L = 202\ \mu m$	$L = 150\ \mu m$	$L = 100\ \mu m$
Width	$w = 10\ \mu m$	$w = 90\ \mu m$	$w = 10\ \mu m$
Thickness	$t = 7\ \mu m$	$t = 7\ \mu m$	$t = 7\ \mu m$
Capacitive gap width	$g = 500\ nm$	$g = 500\ nm$	$g = 500\ nm$
Measurement Conditions			
Polarization voltage	$V_{DC} = 40$ V	$V_{DC} = 40$ V	$V_{DC} = 70$ V
AC voltage	$V_{AC} = 0$ dBm	$V_{AC} = 0$ dBm	$V_{AC} = 0$ dBm
Pressure level	Vacuum pressure: 0.5 Pa	Vacuum pressure: 0.5 Pa	Packaging at: 0.02 Pa
Measurement Results			
Resonance frequency	10.86 MHz	44.66 MHz	20.24 MHz
Quality factor	4400	22,300	50,600
Peak amplitude	2.5 dB	1.2 dB	0.4 dB
Effective mass	1.65×10^{-11} kg	1.1×10^{-10} kg	9.38×10^{-12} kg
Effective stiffness	76.87×10^3 N/m	12.42×10^6 N/m	155.28×10^3 N/m
Electromechanical transduction factor	9.92×10^{-8}	1.49×10^{-6}	1.74×10^{-7}
Motional Parameters			
Motional capacitance	0.13 aF	0.18 aF	0.19 aF
Motional inductance	1676.72 H	49.55 H	309.82 H
Motional resistance	26 MΩ	0.75 MΩ	0.79 MΩ
Feedthrough capacitance	1.24×10^{-15} F	1.86×10^{-14} F	1.24×10^{-15} F

are observed, and the specifications are summarized in Table 4.1. For resonator 1, the frequency characteristics of 10.86 MHz, with a length of one arm of 202 μm, width of 10 μm, thickness of 7 μm, and capacitive gap size of 500 nm are evaluated.

Figure 4.8 presents measured resonant frequency characteristics at pressures of 0.5 Pa and ambient atmosphere under measurement conditions $V_{DC} = 40$ V and $V_{AC} = 0$ dBm. The Q factor of 4400 at 0.5 Pa is larger than that of 1200 in the ambient atmosphere due to the absence of air viscous damping. The calculation of equivalent circuit parameters of resonator 1 based on measured data shows motional capacitance $C_m = 0.13$ aF, motional inductance $L_m = 1676.72$ H, motional resistance $R_m = 26$ MΩ, and peak amplitude of 2.5 dB.

The measured resonant frequency is in good agreement with both FEM simulation and theoretical results, as shown in Figure 4.9. The two-arm-type resonator mainly depends on the length of the arm. When increasing the length of the arm of

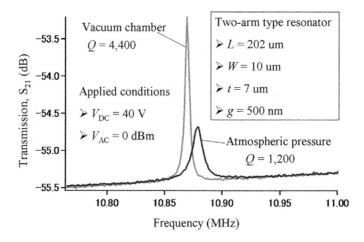

FIGURE 4.8 Frequency response of two-arm-type resonator with pressure chamber and atmospheric pressure both under 0.5 Pa.

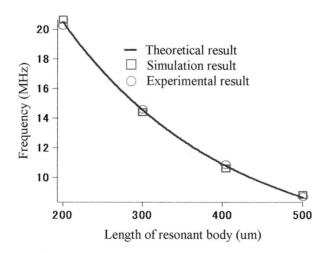

FIGURE 4.9 Resonant frequency versus changing the length of resonant structure.

the resonator, the resonator frequency will decrease. The resonant frequency of the two-arm-type resonator can also be simply calculated with an approximate result, using Equation 4.1 as given:

$$f_0 = \frac{1}{2\pi}\sqrt{\frac{k_{\text{eff}}}{m_{\text{eff}}}} = \frac{1}{2L}\sqrt{\frac{E}{\rho}}, \qquad (4.1)$$

where L is the length of two arms of the resonator, and E and ρ are the Young's modulus and density of the structural material, respectively.

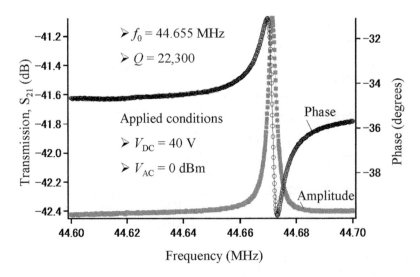

FIGURE 4.10 Transmission and phase response of bar-type silicon resonator at 0.5 Pa pressure in vacuum chamber.

As shown in Figure 4.10, the measurement results of the transmission S_{21} and phase response are indicated for resonator 2 with a length of 150 μm, width of 90 μm, thickness of 7 μm, and capacitive gap width of 500 nm. A resonant peak, which is observed under the conditions $V_{DC} = 40$ V and $V_{AC} = 0$ dBm, is found at 44.66 MHz with the Q factor of 22,300. The calculation of equivalent circuit parameters of resonator 2 shows motional capacitance $C_m = 0.18$ aF, motional inductance $L_m = 49.55$ H, motional resistance $R_m = 0.75$ MΩ, and peak amplitude 1.2 dB.

There is large difference between the motional resistance value of resonator 1 and that of resonator 2 because the effective stiffness constant of resonator 2 ($k_{eff} = 12.42 \times 10^6$ N/m) is much larger than that of resonator 1 ($k_{eff} = 76.87 \times 10^3$ N/m), and also the electromechanical transduction factor η of resonator 1 ($\eta = 9.92 \times 10^{-8}$) is much smaller than that of resonator 2 ($\eta = 1.49 \times 10^{-6}$). The motional resistance should become a much lower value, which will be obtained by reducing the capacitive gap width [13–15]. As an example, for the same structure with a resonant frequency of 44.66 MHz under the same conditions about the DC and AC voltage and assuming no change in Q factor, the motional resistance will be reduced to 19 kΩ by reducing the gap size to 200 nm.

The observed peak amplitude is small in both resonator 1 (2.5 dB) and resonator 2 (1.2 dB) due to the larger capacitive gap size and no amplification of the signal. The high-performance device will be obtained by decreasing the gap size, using an external capacitor for feed-through cancellation or using piezoelectric transduction [15–19].

The resonant peak becomes higher and narrower and its resonant frequency decreases when increasing the polarization voltage V_{DC} due to electrical stiffness (spring softening). A frequency tuning with a shift of 27 kHz is demonstrated for resonator 2 by changing the polarization voltage V_{DC} from 10 V to 40 V, as shown

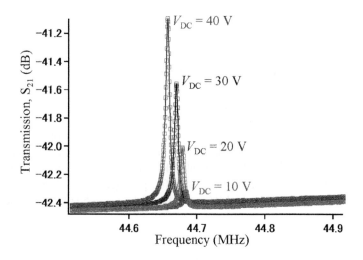

FIGURE 4.11 Changing of the polarization voltage V_{DC}.

in Figure 4.11. When the resonator is deformed from equilibrium by an electrostatic force F_E, a restoring force F_R will be created from the elastic spring stiffness, which acts to bring the resonator back toward equilibrium, as in Figure 4.12.

The electrostatic force that depends on the polarization voltage is acting in the opposite direction from the elastic restoring spring force. The restoring force is effectively reduced, so the structure acts as though it has a reduced spring constant with an increasing polarization voltage V_{DC}. Therefore, the resonant peak shifts to lower frequency if the polarization voltage V_{DC} is increased. The effect of the electrical stiffness caused by the polarization voltage V_{DC} on the resonant frequency of resonator 2 is clearly explained by Equation 4.2:

$$f_{\text{resonator}2} = \frac{1}{2\pi}\sqrt{\frac{k_{\text{eff}}}{m_{\text{eff}}}} = \frac{1}{2\pi}\sqrt{\frac{k_m + k_e}{m_{\text{eff}}}} = \frac{1}{2\pi}\sqrt{\frac{\pi^2 Eh - \dfrac{V_{DC}^2 \varepsilon_0 Lt}{2g^3}}{m_{\text{eff}}}}, \qquad (4.2)$$

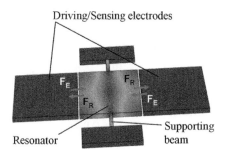

FIGURE 4.12 Electrostatic force and restoring force of bar-type silicon resonator structure.

where $k_m = \pi^2 Et$ and $k_e = -(V_{DC}^2 \varepsilon_0 Lt)/(2g^3)$ are the mechanical and electrical spring constants, respectively [19]. L and t are the length and thickness of the bar-type resonator, respectively.

The measured electrostatic tuning characteristic for the 44.66 MHz bar-type resonator with 500 nm capacitive gaps is around 900 Hz/V, as shown in Figure 4.13. The curve fitting results show a quadratic dependence of the resonant frequency of the polarization voltage V_{DC}. This is in agreement with Equation 4.2.

The resonant characteristics, transmission S_{21} response, and phase response of resonator 3 with a length of one arm of 100 μm, width of 10 μm, thickness of 7 μm, and gap width of 500 nm are shown in Table 4.1 and Figure 4.14, and measured under the condition $V_{DC} = 70$ V and $V_{AC} = 0$ dBm after packaging at 0.02 Pa. The resonant frequency is observed at 20.24 MHz, and the Q factor is 50,600. The theoretical

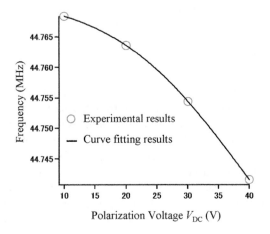

FIGURE 4.13 Frequency tuning characteristic for 44.66 MHz bar-type silicon resonator structure.

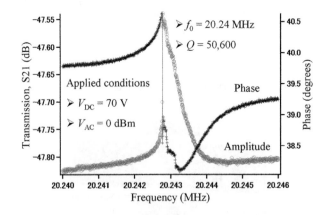

FIGURE 4.14 Transmission and phase response of two-arm-type silicon resonator after packaging at 0.02 Pa.

result of the resonant frequency is calculated to be 20.49 MHz, which is in good agreement with the experiment result and simulation result (20.65 MHz).

The resonant frequency of the resonator structure shifts when changing the polarization voltage V_{DC} shown in Figure 4.15. This characteristic of resonator 3 is similar to that of resonator 2. An electrostatic force F_E and a restoring force F_R are shown in Figure 4.16. The effect of the electrical stiffness caused by the polarization voltage on the resonant frequency of resonator 3 is given by

$$f_{resonator3} = \frac{1}{2\pi}\sqrt{\frac{k_{eff}}{m_{eff}}} = \frac{1}{2\pi}\sqrt{\frac{k_m + k_e}{m_{eff}}} = \frac{1}{2\pi}\sqrt{\frac{\frac{\pi^2}{8}\frac{Ewt}{L} - \frac{V_{DC}^2 \varepsilon_0 wt}{2g^3}}{m_{eff}}}, \quad (4.3)$$

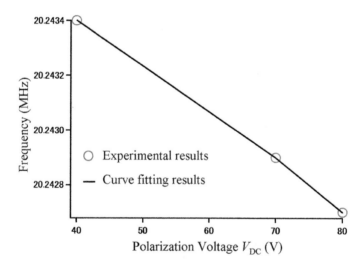

FIGURE 4.15   Frequency tuning characteristic for 20.24 MHz two-arm-type silicon resonator structure.

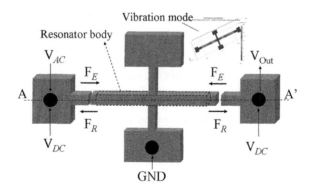

FIGURE 4.16   Electrostatic force and restoring force of two-arm-type silicon resonator structure.

where $k_m = (\pi^2 Ewt) / (8L)$ and k_e are the mechanical and electrical spring constant, respectively.

Figure 4.15 shows the measured electrostatic frequency tuning characteristic for the two-arm-type resonator after the packaging process, indicating a frequency tuning range of 700 Hz, by changing the polarization voltage from 40 V to 80 V. The measured electrical tuning slope is 17.5 Hz/V, showing an effective capacitive gap size of 500 nm after packaging.

4.6 SUMMARY

A new device concept for packaging using the LTCC substrate with an Au metal feed-through is demonstrated in this chapter. The fabrication and packaging processes of micromechanical silicon resonators on the LTCC substrate are successfully performed. The silicon resonator structures are formed on the LTCC substrate using the anodic bonding of silicon and LTCC for electrical interconnections. Then the resonator structures are packaged hermetically by the second anodic bonding of silicon and Tempax glass for encapsulation. The characterization of silicon resonators is evaluated in the vacuum chamber and ambient atmospheric pressure. It is shown that the vacuum is indispensable to obtain the high Q factor for the silicon resonators. The measured resonant frequency of the packaged device is 20.24 MHz and the high Q factor of 50,600 is observed without any kind of amplification.

REFERENCES

1. Esashi, M., Wafer level packaging of MEMS, *J. Micromech. Microeng.*, **18**, 073001, 2008.
2. Pourkamali, S., Ayazi, F., Wafer-level encapsulation and sealing of electrostatic HARPSS transducers, In: *Proceedings of IEEE Sensor*, 49–52, 2007.
3. Fang, J., Fu, J., Ayazi, F., Meal organic thin film encapsulation for MEMS, *J. Micromech. Microeng.*, **18**, 105002, 2008.
4. Li, X., Abe, T., Liu, Y., Esashi, M., Fabrication of high density electrical feed through by deep reactive ion etching of Pyrex glass, *J. Microelectromech. Syst.*, **11**, 625–630, 2002.
5. Nakamura, K., Takayanagi, F., Moro, Y., Sanpei, H., Onozawa, M., Esashi, M., Development of RF MEMS switch, In: *Proceedings of Advantest Technical Report*, **22**, 9–15, 2004.
6. Tanaka, S., Matruzaki, S., Mohri, M., Okada, A., Fukushi, H., Esashi, M., Wafer level hermetic packaging technology for MEMS using anodically bondable LTCC wafer, In: *Proceedings of Micro Electro Mechanical Systems*, 376–379, 2011.
7. Tanaka, S., Mohri, M., Okada, A., Fukushi, H., Esashi, M., Versatile wafer-level hermetic packaging technology using anodically bondable LTCC wafer with compliant porous gold bumps spontaneously formed in wet etched cavities, In: *Proceedings of Micro Electro Mechanical Systems*, 369–372, 2012.
8. Lin, Y.C., Wang, W.S., Chen, L.Y., Chen, M.W., Gessner, T., Esashi, M., Anodically bondable LTCC substrate with novel nano-structured electrical interconnection for MEMS packaging, In: *Proceedings of Solid-State Sensors, Actuators and Microsystems Conference*, 2351–2354, 2011.

9. Toan, N.V., Miyashita, H., Toda, M., Kawai, Y., Ono, T., Fabrication of an hermetically packaged silicon resonator on LTCC substrate, *Microsyst. Technol.*, **19**, 1165–1175, 2013.
10. Toan, N.V., Toda, M., Kawai, Y., Ono, T., A capacitive silicon resonator with a movable electrode structure for gap width reduction, *J. Micromech. Microeng.*, **24**, 025006, 2014.
11. Toan, N.V., Kubota, T., Sekhar, H., Samukawa, S., Ono, T., Mechanical quality factor enhancement in silicon micromechanical resonator by low-damage process using neutral beam etching technology, *J. Micromech. Microeng.*, **24**, 085005, 2014.
12. Toan, N.V., Toda, M., Kawai, Y., Ono, T., A long bar type silicon resonator with a high quality factor, *IEEJ Trans. Sens. Micromachines*, **134**, 26–31, 2014.
13. van Beek, J.T.M., Puers, R., A review of MEMS ocsillators for frequency reference and timing applications, *J. Micromech. Microeng.*, **22**, 013001, 2012.
14. Nguyen, C.T.C., MEMS technology for timing and frequency control, *IEEE Trans. Inltrason. Ferroelectr. Freq. Control*, **54**, 251–270, 2007.
15. Pourkamali, S., Ho, G.K., Ayazi, F., Low impedance VHF and UHF capacitive silicon bulk acoustic wave resonators – Part I: Concept and fabrication, *IEEE Trans. Electron Devices*, **54**, 2017–2023, 2007.
16. Pourkamali, S., Ho, G.K., Ayazi, F., Low impedance VHF and UHF capacitive silicon bulk acoustic wave resonators – Part II: Measurement and characterization, *IEEE Trans. Electron Devices*, **54**, 2024–2030, 2007.
17. Lee, J.E.-Y., Seshia, A.A., 5.4-MHz single-crystal silicon wine glass mode disk resonator with quality factor of 2 million, *Sensors and Actuators A: Physical*, **156**, 28–35, 2009.
18. Samarao, A.K., Ayazi, F., Combined capacitive and piezoelectric transduction for high performance silicon microresonators, In: *Proceedings of Micro Electro Mechanical Systems (MEMS 2011)*, 169–172, 2011.
19. Kaajakari, V., Mattila, T., Oja, A., Seppa, H., Nonlinear limits for single crystal silicon microresonators, *J. Microelectromech. Syst.*, **13**, 715–724, 2004.

5 A Long-Bar-Type Capacitive Silicon Resonator with a High Quality Factor

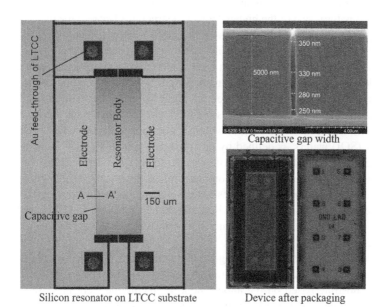

Silicon resonator on LTCC substrate   Device after packaging

5.1 INTRODUCTION

Single-crystal silicon is usually used as a structural material for mechanical resonators because of a low internal frictional loss and consequently high mechanical quality (Q) factor [1–13]. A high Q factor is not only desired to reduce the motional resistance but also helps to reduce phase noise performance of oscillators. Besides the choice of the structural material, the Q factor is also affected by the topology of the device structure. Methods for increasing the Q factor of silicon resonators have been reported [8, 9]. A study of supporting beam geometries of a square-type resonator for reducing the anchor loss and accordingly increasing the Q factor was performed [8]. A bulk-mode resonance of single-crystal silicon microresonator structures with a deep cavity under the resonator to avoid the parasitic capacitance between the device and the substrate exhibits a high Q factor [9]. In this chapter, a simple method to obtain a high Q factor and a low motional resistance by changing the dimensions of the resonant structure is proposed. The Q factor, which is expressed as a function of a resonant frequency f_0 and a characteristic resonator dimension D, was presented by van Beek and Puers [1], as given by

$$Q = 2\pi \frac{f_0 m}{\gamma} =\sim 2\pi f_0 D,$$ (5.1)

where m is the effective mass and γ is the damping coefficient.

Equation 5.1 shows that the Q factor is a function of the resonator dimension. Even if the resonators of different structures have the same resonant frequency, the Q factor is variable by the different dimensions. Thus, if the resonant frequency is kept constant and the value of D is increased, the Q factor becomes larger. The bar-type silicon resonator is one of the highest potential structures because the dimensions can be changed by the length or thickness of the resonant body without degradation of its resonant frequency. However, if the thickness of the resonator structure is increased, a high aspect-ratio etching silicon must be considered to achieve the narrow capacitive gap width. Therefore, changing resonator length is a better solution to increase the Q factor. Increasing the length of the structure also increases the transduction area. Therefore, a high Q factor and the desired motional resistance value can be achieved.

The silicon resonator structure, which is commonly fabricated from silicon on insulator (SOI) wafers, is defined by deep reactive ion etching (RIE) of the device layer and released by etching the sacrificial layer using hydrofluoric (HF) acid [10–13]. This fabrication process is slightly complex and becomes more challenging for long-type structures due to stiction issues for the resonant structure during HF release. Moreover, the parasitic capacitance between the device layer and the handling silicon layer, which affected the resonator performance, is very large since the sacrificial layer is very thin.

A fabrication process for the long-bar-type silicon resonators to avoid the aforementioned issues and a packaging process using a low temperature co-fired ceramic (LTCC) substrate based on the anodic bonding technique were proposed and presented in Chapter 4. Long-bar-type resonators were fabricated and hermetically packaged. The resonant characteristics before and after the packaging process were observed. The vacuum level of a hermetically packaged device was also evaluated. The detailed information is as follows.

5.2   DEVICE STRUCTURE AND FABRICATED RESULTS

Resonator structures in this section are the bar-type resonators that are excited in the horizontal width-extensional mode at the resonant frequency. The finite element method (FEM) simulation result for the vibration shape is shown in Chapter 2, Figure 2.6b. The resonator's parameters are shown in Table 5.1.

The fabrication process of the long-bar-type silicon resonators is the same as mentioned in Chapter 4. The fabricated results are shown in Figure 5.1. The structure of the silicon resonators is defined by deep RIE of the top silicon device layer of an SOI wafer, which consists of a 5 μm-thick top silicon layer, 1 μm-thick oxide layer, and

TABLE 5.1

Summarized Parameters of the Bar-Type Capacitive Silicon Resonators with and without the Packaging Process

	Unpackaged Device			Packaged Device
Resonator Parameters				
Length of resonant body (µm)	$L=500$	$L=1000$	$L=1500$	$L=1500$
Width of resonant body (µm)	$W=440$	$W=440$	$W=440$	$W=440$
Thickness of resonant body (µm)	$t=5$	$t=5$	$t=5$	$t=5$
Capacitive gap (nm)	$g=400$	$g=300$	$g=300$	$g=300$
Applied Conditions				
V_{DC} (V)	$V_{DC}=25$	$V_{DC}=10$	$V_{DC}=10$	$V_{DC}=10$
V_{AC} (dBm)	$V_{AC}=0$	$V_{AC}=0$	$V_{AC}=0$	$V_{AC}=0$
Theoretical Calculation				
Effective mass (10^{-9} kg)	$m_{eff}=1.28$			$m_{eff}=3.84$
Effective stiffness (10^6 Nm^{-1})	$k_{eff}=4.75$			$k_{eff}=14.3$
Resonant frequency (MHz)	$f_0=9.7$			$f_0=9.7$
Electromechanical transduction (10^{-6})	$\eta=2.46$			$\eta=7.38$
Results				
Resonance frequency (MHz)	$f_0=9.65$	$f_0=9.68$	$f_0=9.69$	$f_0=9.69$
Quality factor	$Q=64,000$	$Q=148,000$	$Q=368,000$	$Q=341,000$
Insertion loss (dB)	$IL=-71$	$IL=-62$	$IL=-56$	$IL=-56$
Motional resistance (kΩ)	$R_m=100$	$R_m=43.3$	$R_m=11.7$	$R_m=12.5$
Motional capacitance (aF)	$C_m=2.5$	$C_m=2.6$	$C_m=3.8$	$C_m=3.8$
Motional inductance (H)	$L_m=107$	$L_m=10.6$	$L_m=70$	$L_m=70$
Feedthrough capacitance (pF)	$C_f=0.06$	$C_f=0.15$	$C_f=0.22$	$C_f=0.22$

a 475 µm-thick silicon handling layer. Then, the patterned top Si layer is transferred onto an LTCC substrate. The resonator is hermetically sealed by the anodic bonding technique. Figure 5.1a shows the micrograph of the bar-type capacitive silicon resonator formed on LTCC substrate with the capacitive gap width of approximately 300 nm (Figure 5.1a). The front side and back side of the fabricated device after the sealing process are shown in Figure 5.1c.

5.3 PARASITIC CAPACITANCE CANCELLATION

Measurement setup for the characterization of the capacitive silicon resonators was illustrated in Chapter 4. In this section, the cancellation of the effects of the parasitic elements is presented in detail. The parasitic capacitances come from several sources such as feed-through capacitances C_f, pad capacitances C_p, and other parasitic capacitances associated with a printed circuit board. Although the parasitic capacitance value is quite small, for high-frequency applications the electrical path through

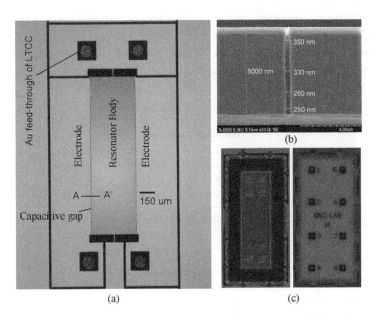

FIGURE 5.1 Fabricated results. (a) Micrograph of silicon resonator on LTCC substrate. (b) A–A' cross section. (c) Front side and back side of the device after packaging process.

these capacitances is significant and could potentially obscure the resonant signal. Therefore, one of the biggest problems still facing the capacitive silicon resonators for oscillators or other applications is a small resonant signal. The amplifier circuit is commonly used to obtain larger resonant signal [13, 14]. Another method is to use capacitive compensation. This capacitance can suppress the parasitic elements [15]. Here, the simple method is to observe the resonant signal only for exactly evaluating the Q factor, and the motional resistance of the silicon resonators is presented. The cancellation of the effects of the parasitic elements is based on a function of the vector network analyzer, namely, Agilent E5071B.

The motional components of the capacitive silicon resonators are given by the function of the applied polarization voltage V_{DC}. The motional resistance and motional inductance become a large value, and the motional capacitance value becomes very small when the polarization voltage becomes zero ($V_{DC} \rightarrow 0$). Thus, there is almost no current through the motional components in this case. The measured transmission of the resonator without applied polarization voltage ($V_{DC} = 0$) is influenced by the parasitic elements I_p only while that of the resonator, measured as a function of applied polarization voltage ($V_{DC} \neq 0$), consists of both the motional I_m and parasitic I_p components. In this manner, it is possible to extract the motional component by subtracting the measured results under the aforementioned two different conditions (with and without the polarization voltage V_{DC}). The accurate measurement of the motional component has been achieved using this method, as follows

$$I_{\text{output}} = I_{\text{with } V_{DC}} - I_{\text{without } V_{DC}} = \left(I_m + I_p\right)_{\text{with } V_{DC}} - \left(I_p\right)_{\text{without } V_{DC}} = I_m, \qquad (5.2)$$

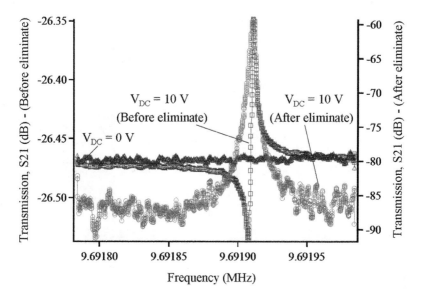

FIGURE 5.2 Cancellation of parasitic elements.

Figure 5.2 presents the typical amplitude signal of the capacitive silicon resonators that are measured in the vacuum environment (vacuum chamber). A clear resonant peak is observed after performing the cancellation of the parasitic capacitance elements.

5.3.1 MEASUREMENT RESULTS

The measurement results of the fabricated bar-type capacitive silicon resonators before and after packaging will be presented and discussed in this section. Before the packaging process, the resonators are placed inside a vacuum chamber with coaxial feed-through cables. The resonant characteristics of the bar-type resonators are observed, and the specifications are summarized in Table 5.1. As shown in Figure 5.3, the measurement result of the transmission S_{21} is indicated for the bar-type resonator with a length of 1500 μm, the width of 440 μm, the thickness of 5 μm, and the capacitive gap size of around 300 nm. A resonant peak, which is observed under measurement conditions V_{DC} of 10 V and V_{AC} of 0 dBm, is found at 9.69 MHz with an ultra-high Q factor of 368,000.

The theoretical calculation of the resonant frequency is 9.7 MHz, which is in good agreement with the experiment result (9.69 MHz) and the FEM simulation result (9.51 MHz). The calculation of the equivalent circuit parameters of the silicon resonator based on the measured data shows the motional capacitance $C_m = 3.8$ aF, the motional inductance $L_m = 70$ H, the motional resistance $R_m = 11.7$ kΩ, and the feed-through capacitance $C_f = 0.22$ pF.

The theoretical analysis, the FEM simulations, and the experimental results show that the resonant frequency of the silicon bar-type resonator is determined by the width of the resonant body mainly. Therefore, the operating frequency can

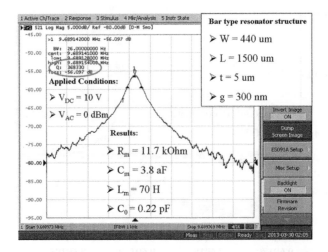

FIGURE 5.3 Transmission S_{21} signal of the bar-type resonator before packaging.

be maintained while extending the length or thickness of the structure to provide a large transduction area for obtaining a high Q factor and low motional resistance, as discussed in Section 5.1. The experimental results indicate that the long-bar-type resonator body has a higher Q factor and lower motional resistance than those of a shorter resonator body, as shown in Table 5.1. The obtained experimental results are in agreement with Equation 5.1. The resonant frequency of the bar-type resonators can also be simply calculated with an approximate equation as follows:

$$f_0 = \frac{1}{2\pi}\sqrt{\frac{k_{\mathrm{eff}}}{m_{\mathrm{eff}}}} = \frac{1}{2W}\sqrt{\frac{E}{\rho}}, \tag{5.3}$$

where $k_{\mathrm{eff}} = \pi^2 E L t / 2W$, $m_{\mathrm{eff}} = \rho L W t / 2$, W is the width of bar type resonator, and E and ρ are Young's modulus and density of the structural material, respectively. Equation 5.3 clearly shows that the resonant frequency of the bar-type resonator depends on the width of the resonant structure only; consequently, a wide range of the resonant frequency can be simultaneously fabricated on the same substrate.

Figure 5.4 shows the relationship between the resonant frequency and the width of the bar-type structure. An increase in the width of the bar-type structure results in the decrease of the resonant frequency. The relationship between the Q factor and the pressure ambient is measured, and the result is shown in Figure 5.5. The Q factor of long-bar-type silicon resonators was obtained for different pressures, from atmosphere pressure ($\sim 10^5$ Pa) to vacuum pressure ($\sim 10^{-2}$ Pa). The figure shows the effects of the viscous damping on the Q factor when the pressure ambient is changed. The Q factor, calculated by the network analyzer based on its associated resonant peak, is 3000 in air and 368,000 at 0.01 Pa. Resonant spectra in the air, the vacuum chamber, and the hermetically packaged device are compared as shown in Figure 5.6. The silicon resonator was packaged at 0.008 Pa; however, the Q factor of 341,000 of the hermetically packaged device is smaller than that of the silicon resonator at

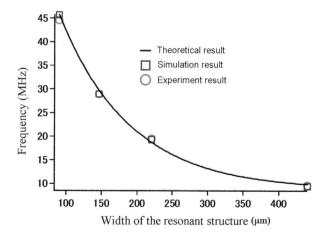

FIGURE 5.4   Width dependence on resonant frequency of bar-type resonator.

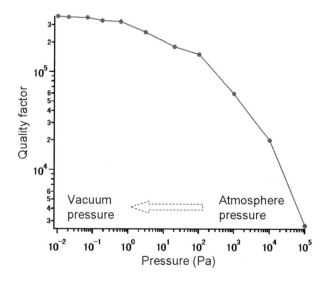

FIGURE 5.5   Relationship between the Q factor and the ambient pressure.

0.01 Pa in the vacuum chamber. One of the main reasons is due to the outgassing in the bonding process. From the evaluation based on the Q factor, the pressure of the sealed cavity is estimated to be around 0.1 Pa to 10 Pa.

Figure 5.7 presents a comparison of the Q factor between short- and long-term operations of the packaged silicon resonator device. The Q factor drift with the operation time is observed. As shown in Figure 5.7, the Q value is decreased two times over 6 months (long-term operation). The possible reasons are the leakage of the hermetically packaged or the outgassing of the metal in the cavity that makes the pressure in the packaged cavity increase. The pressure of the sealed cavity is estimated

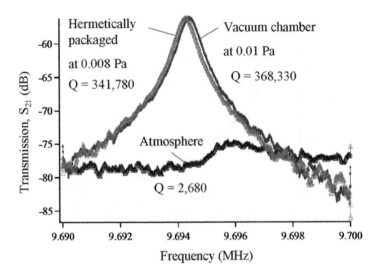

FIGURE 5.6   Comparison of the performance of the resonator with and without packaging process.

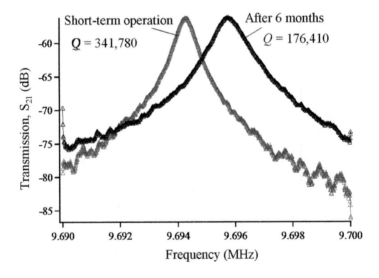

FIGURE 5.7   Short-term operation and long-term operation of packaged device.

to be around 100 to 300 Pa based on the evaluation of Q factor. The possible solution to solve the problems is the use of getter materials.

5.4   SUMMARY

The long-bar-type silicon resonator with a high Q factor and low motional resistance is demonstrated. The experimental results show the effects of the length of the resonant body on the Q factor. Additionally, the device is hermetically packaged using an LTCC substrate based on the anodic bonding technique. The resonant characteristics before and after the packaging process are evaluated. The resonator is excited in the extensional mode at a resonant frequency of 9.69 MHz. The value of the Q factor measured for this device is 368,000 at a vacuum chamber pressure of 0.01 Pa and 341,000 after the packaging process. The dependence of the Q factor on the operating pressure is measured in order to evaluate the vacuum level of a hermetically packaged device.

REFERENCES

1. van Beek, J.T.M., Puers, R., A review of MEMS oscillators for frequency reference and timing applications, *Journal of Micromechanics and Microengineering*, **22**, 013001, 2012.
2. Nguyen, C.T.C., MEMS technology for timing and frequency control, *IEEE Transactions on Ultrasonics, Ferroelectrics, and Frequency Control*, **54**, 251–270, 2007.
3. Ayazi, F., MEMS for integrated timing and spectral processing, *IEEE 2009 Custom Integrated Circuits Conference (CICC)*, San Jose, CA, 65–72, 2009.
4. Toan, N.V., Miyashita, H., Toda, M., Kawai, Y., Ono, T., Fabrication and packaging process of silicon resonator capable of the integration of LSI for application of timing device, *The 26th IEEE International Conference on Micro Electro Mechanical Systems*, Taipei, Taiwan, 377–380, 2013.
5. Toan, N.V., Miyashita, H., Toda, M., Kawai, Y., Ono, T., Fabrication of an hermetically packaged silicon resonator on LTCC substrate, *Microsystem Technologies*, **19**, 1165–1175, 2013.
6. Akgul, M., Kim, B., Hung, L.W., Lin, Y., Li, W.C., Huang, W.L., Gurin, I., Borna, A., Nguyen, C.T.C., Oscillator far from carrier phase noise reduction via nano scale gap tuning of micromechanical resonator, *The 15th International Conference on Solid State Sensors, Actuators and Microsystems*, Denver, CO, 798–801, 2009.
7. Lee, S., Nguyen, C.T.C., Mechanical coupled micromechanical resonator arrays for improved phase noise, *Proceedings of IEEE International Conference on Ultrasonics, Ferroelectrics and Frequency Controls*, 280–286, 2004.
8. Lee, J.E.Y., Yan, J., Seshia, A.A., Study of lateral mode SOI-MEMS resonators for reduced anchor loss, *Journal of Micromechanics and Microengineering*, **21**, 045011, 2011.
9. Wu, G., Xu, D., Xiong, B., Wang, Y., A high performance bulk mode single crystal silicon microresonator based on a cavity-SOI wafer, *Journal of Micromechanics and Microengineering*, **22**, 025020, 2012.
10. Pourkamali, S., Ho, G.K., Ayazi, F., Low-impedance VHF and UHF capacitive silicon bulk acoustic wave resonators – Part I: Concept and fabrication, *IEEE Transactions on Electron Devices*, **54**, 2017–2023, 2007.
11. Ho, G.K., Perng, J.K., Ayazi, F., Micromechanical IBARs: Modeling and process compensation, *Journal of Microelectromechanical Systems*, **19**, 516–525, 2010.
12. Ho, G.K., Sudaresan, K., Pourkamali, S., Ayazi, F., Low motional impedance highly tunable I^2 resonator for temperature compensated reference oscillators, *The 18th IEEE International Conference on MEMS*, Miami Beach, FL, 116–120, 2005.
13. Mattila, T., Kiihamäki, J., Lamminmäki, T., Jaakkola, O., Rantakari, P., Oja, A., Seppä, H., Kattelus, H., Tittonen, I., A 12 MHz micromechanical bulk acoustic mode oscillator, *Sensors and Actuators A: Physical*, **101**, 1–9, 2002.

14. Kim, B., Hopcroft, M.A., Candler, R.N., Jha, C.M., Agarwal, M., Melamud, R., Chandorkar, S.A., Yama, G., Kenny, T.W., Temperature dependence of quality factor in MEMS resonators, *Journal of Microelectromechanical Systems*, **17**, 755–766, 2008.
15. Lee, J.E.-Y., Seshia, A.A., Parasitic feed-through cancellation techniques for enhanced electrical characterization of electrostatic microresonators, *Sensors and Actuators A: Physical*, **156**, 36–42, 2009.

6 Capacitive Silicon Resonators Using Neutral Beam Etching Technology

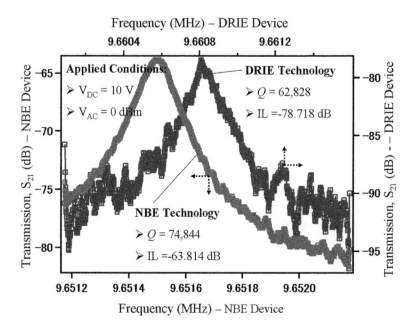

6.1 INTRODUCTION

Silicon on insulator (SOI) wafers are widely used for resonator fabrication [1–7] because single-crystalline silicon has a low internal frictional loss, thus, silicon resonators consequently have a high mechanical Q factor. A high Q factor is not only desired to reduce the motional resistance but also helps to reduce the phase noise of oscillators [1]. Besides the choice of the structural material, the Q factor is also affected by many other factors [1]. One of them is surface defects that make the energy dissipation larger. Therefore, the Q factor gets smaller. The formed structures on the silicon device layer are commonly performed by deep reactive ion etching (DRIE) using the Bosch process. Disadvantages of DRIE technology are charge-induced damage or damage due to ultraviolet/vacuum ultraviolet (UV/VUV) light from the high-density plasma during the etching process [8–11]. Additionally, DRIE also creates rough surfaces on the sidewalls, which is referred to as scallops or ripples. However, neutral beam etching (NBE) technology [12–14] does not possess charge, and UV/VUV light is blocked by an aperture during etching. NBE technology can achieve etching with almost no damage and atomically flat silicon surfaces; thus, the NBE process is appropriate for micro/nano fabrication.

In this chapter, NBE technology is investigated to find a highly anisotropic etching shape and as a fabrication method for silicon micromechanical resonators to a obtain narrow capacitive gap and smooth etching surface for small motional resistance, low insertion loss, and high Q factor. The devices fabricated by both DRIE and NBE are evaluated and compared. Moreover, a comparison of damages between NBE and DRIE in terms of mechanical Q factors is discussed. This work reveals a new approach to fabricate silicon micromechanical resonators using NBE.

6.2   NEUTRAL BEAM ETCHING TECHNOLOGY

6.2.1   SAMPLE PREPARATION

The sample preparation process for testing NBE is started from an SOI wafer as shown in Figure 6.1a, which consists of a 5 µm-thick top silicon device layer, 1 µm-thick oxide layer, and 475 µm-thick silicon handling layer. A 1 µm-thick SiO_2 layer

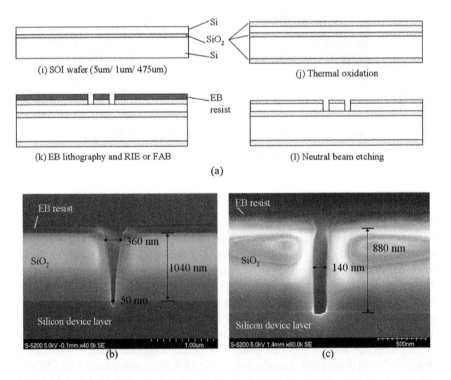

FIGURE 6.1   (a) Si etching process for nanoscale patterning. (b) FAB etching result for SiO_2 mask pattern (mask 1). (c) RIE etching result for SiO_2 mask pattern (mask 2).

is grown on the SOI by the wet thermal oxidation (1100°C and 3.5 hours) process for the etching mask of silicon. To make nanotrenches, electron beam lithography (EBL) and high-resolution positive electron beam resist (ZEP 520A) are used. Then, the SiO_2 mask layer is etched by two techniques. One of the techniques is fast atom beam (FAB) using CHF_3 gas. During etching, the anode–cathode voltage is 2.5kV and the discharge current is 25 mA. The FAB etching result shows a taper trench shape (mask 1), as shown in the cross-sectional image in Figure 6.1b. Another technique is reactive ion etching (RIE) using a gas mixture of CHF_3 and Ar gases at a power of 120W and a chamber pressure of 5 Pa as shown in the vertical shape (mask 2) of Figure 6.1c. Masking patterns with two kinds of cross-sectional shapes (taper and vertical shapes) are prepared for testing NBE.

6.2.2 NEUTRAL BEAM ETCHING APPARATUS

The schematic of the NBE system used in our research is shown in Figure 6.2 [13]. A 200 mm diameter neutral beam source is employed to etch the silicon device layer. This system consists of an inductively coupled plasma (ICP) chamber and a sample chamber separated by a high-aspect-ratio carbon aperture. The ICP is generated using a radio frequency (RF) of 13.56 MHz. The RF input power is 2 kW, which is time-modulated at 10 kHz with a duty ratio of 50% (50 μs ON/50 μs OFF). The neutral beam is produced from a gas mixture of pure Cl_2, F_2, and O_2 gases. The ICP contains positive and negative ions and produces high-energy UV/VUV photons. Due to the charge transfer process, ions are neutralized and high-energy UV photons are absorbed by carbon aperture, the only neutral particles participating in the etching process. Therefore, NBE is able to achieve ultra-low damage etching, and a flat silicon sidewall surface can be obtained.

Etching depth, shape, and rate are dependent on several factors, such as the neutralization mechanism [15, 16], bias voltage applied to the apparatus [17], and mask shape. In this chapter, the NBE etching conditions are as follows: A gas mixture of Cl_2 (48 sccm), F_2 (3 sccm), and O_2 (2.4 sccm) is used for etching. The bias power is 900 V and ICP power is 2000 W.

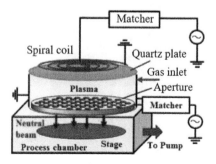

FIGURE 6.2 Schematic of neutral beam etching apparatus.

6.2.3 NBE RESULTS AND DISCUSSION

In this section, we discuss the experimental results of NBE on 5 μm-thick silicon with different mask shapes. Etching depth, etching rate, and shapes of NBE are also investigated by changing etching times. Finally, a comparison of etching results between NBE and DRIE is presented.

First, NBE is performed by using mask 1 (tapered mask) as mentioned in Section 6.1. The SEM image of the cross-sectional shape of the etched profile is shown in Figure 6.3. It is shown that a 5 μm-deep trench is obtained in the silicon device layer of the SOI wafer. The gap widths of the top and center positions of the trench are around 570 nm and 330 nm, respectively. The calculated average value of the gap width is around 450 nm.

Next, NBE is performed by using mask 2 for 60, 120, 180, 240, and 300 minutes, and those etched profiles are shown in Figure 6.4a–e, respectively. Thus, a 5μm-deep trench pattern having smooth sidewalls with a gap width of 230 nm (average value of gap width) is achieved by the NBE process using a vertical mask shape, as shown in Figure 6.4e. Therefore, the etching profile of NBE using a vertical mask shape is more vertical and narrow than those etched using the tapered mask shape in the comparison of etching results between Figures 6.3 and 6.4e. The etching depth of silicon is increased with increasing etching time, but the selectivity between silicon and silicon dioxide decreases. Etching time versus etching depth of silicon and Si/SiO$_2$ selectivity are shown in Figure 6.5.

The silicon device layer of the SOI wafer is also etched by DRIE based on the Bosch process [18–20]. SF$_6$ and C$_4$F$_8$ gases are used for etching and passivation, respectively. This process creates scallops, but the size of the scallop can be reduced to be around 30 nm using short etching and passivation cycles (etching time 4 sec; passivation time 3 sec). A vertical sidewall and small gap width of around 400 nm are obtained as shown in Figure 6.6. Consequently, a gap width fabricated by NBE is smaller than that fabricated by DRIE in the comparison of etching results between Figures 6.4e and 6.6. Additionally, DRIE exhibits scallops of around 30 nm as shown in Figure 6.6, while NBE has a smooth surface (Figure 6.4e). The details of comparison of DRIE and NBE on the etching of 5 μm-thick silicon are summarized in Table 6.1.

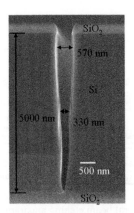

FIGURE 6.3   NBE etching shape with patterned mask by FAB.

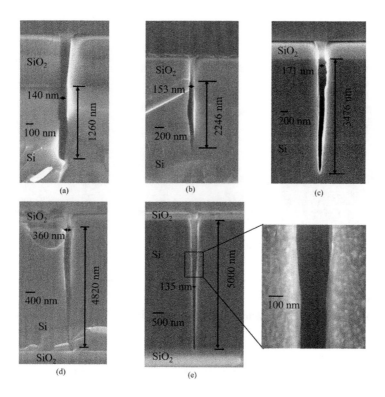

FIGURE 6.4 Cross-sectional images of etched trench by NBE at (a) 60 minutes etching, (b) 120 minutes etching, (c) 180 minutes etching, (d) 240 minutes etching, and (e) 300 minutes etching.

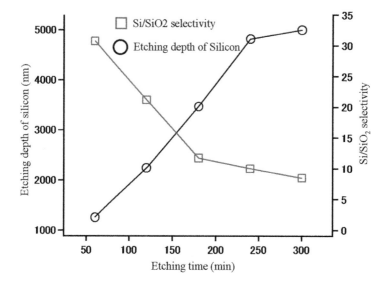

FIGURE 6.5 Time versus etching depth of silicon and Si/SiO_2 selectivity.

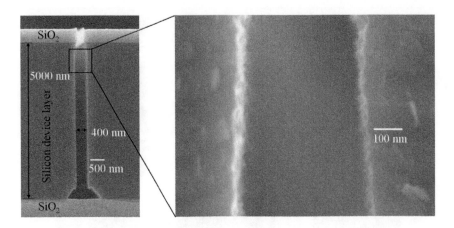

FIGURE 6.6   Etching shape of DRIE process.

TABLE 6.1
Comparison of DRIE and NBE on the Etching of 5 μm-Thick Silicon

	DRIE	NBE
Etching Conditions	Etching material: Si	Etching material: Si
	Mask material: SiO$_2$	Mask material: SiO$_2$
	Gas:	Gas:
	• SF$_6$ – 80 sccm	• C$_{12}$ – 48 sccm
	• C$_4$F$_8$ – 50 sccm	• F$_2$ – 3 sccm
	Etching time: 3 sec	• O$_2$ – 2.4 sccm
	Passivation time: 4 sec	BIAS power: 900 V
	Coil generator: 350 W	ICP power: 2000 W
	Platen generator:	
	• 0 W (passivation process)	
	• 15 W (etching process)	
Etching rate of silicon	125 nm/minutes	17 nm/minutes
Etching rate of SiO$_2$	10 nm/minutes	2 nm/minutes
Selectivity	12.5	8.5
Etching profile	90°	88°
Etching surfaces	Scallop ~30 nm	Smooth

6.3   EXPERIMENTAL METHODOLOGY

Silicon resonator structures are fabricated by following the fabrication process shown in Figure 6.7. The fabrication process details are as follows.

The fabrication process is started from a SOI wafer, which consists of a 5 μm-thick top silicon layer, 1 μm-thick oxide layer, and 475 μm-thick silicon handling layer (Figure 6.7a). Around 1 μm-thick SiO$_2$ is formed on the SOI by wet thermal oxidation for the etching mask of silicon. EBL is performed, which provides a pattern of

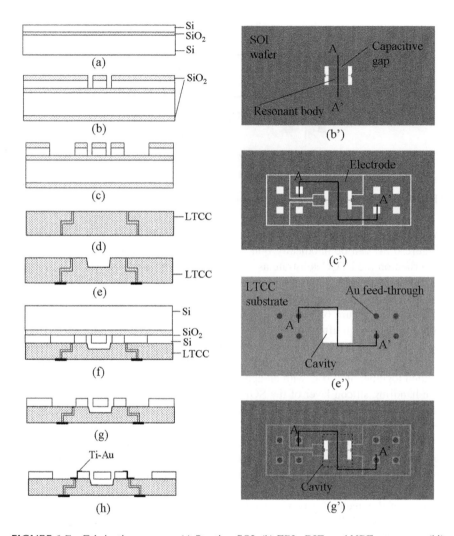

FIGURE 6.7 Fabrication process. (a) Starting SOI. (b) EBL, RIE, and NBE processes. (b')
Top view of the pattern after EBL and NBE processes. (c) Lithography and DRIE processes.
(c') Top view of the pattern after lithography and DRIE processes. (d) Starting LTCC. (e)
Contact pads and etching of LTCC. (e') Top view of the pattern after etching of LTCC. (f)
Anodic bonding. (g) Silicon handling layer and SiO_2 removal. (g') Top view of the pattern
after silicon handling layer and SiO_2 removal. (h) Electrical interconnections.

silicon resonators with narrow capacitive gap width. Then, the SiO_2 layer is etched
by using RIE, as previously described. The capacitive gaps are formed by using NBE
(Figure 6.7b). Figure 6.7b' presents a top view of the pattern after the NBE step. Next,
the resonator structure is created by standard lithography and DRIE (Figure 6.7c),
and the top view of the pattern after lithography and DRIE is shown in Figure 6.7c'.

Cr–Au contact pads with thicknesses of 20 nm and 300 nm, respectively, are
formed by using the wet etching process with photoresist mask on the back side of

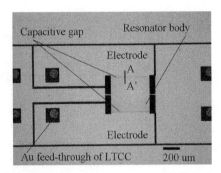

FIGURE 6.8 Micrograph of a silicon resonator formed on an LTCC substrate.

the low temperature co-fired ceramic (LTCC) substrate (Nikko Co.). LTCC substrate (Figure 6.7d) can be anodically bonded to silicon. After a dry film photoresist is patterned on the LTCC substrate as a mask for sandblasting, the LTCC substrate is sandblasted using Al_2O_3 particles to form approximately 100 μm-deep cavities at the underlining position of the resonant body for making freestanding structures (Figure 6.7e). Figure 6.7e' shows the top view of the LTCC substrate with the Au metal feed-through after sandblasting. In order to transfer the patterned silicon on the SOI onto the LTCC substrate, anodic bonding is performed. The patterned SOI wafer (Figure 6.7c) and the patterned LTCC substrate (Figure 6.7e) are aligned and bonded together at 400°C with an applied voltage of 800 V for 5 min (Figure 6.7f). The handling silicon layer of the SOI wafer is etched out by plasma etching using SF_6 gas, and the buried SiO_2 layer is removed by a fast atom beam of CHF_3 gas (Figure 6.7g). The top view of this step is shown in Figure 6.7g'. Then, Ti–Au films with thicknesses of 20 nm and 300 nm, respectively, are partly deposited by a sputtering process via a stencil mask for electrical connections between silicon electrodes and the Au metal feed-through of the LTCC substrate (Figure 6.7h).

Silicon resonators are also successfully fabricated using DRIE. The fabrication process of this method was presented in detail in Chapter 4. The silicon resonator formed on the LTCC substrate is shown in Figure 6.8. The capacitive gap width of the 5 μm-thick devices fabricated by DRIE is around 400 nm as shown in Figure 6.6, while that of the same thick device fabricated by NBE is around 230 nm (average value of the capacitive gap width) (Figure 6.4e). The gap width obtained by NBE is smaller than that obtained by DRIE. Additionally, the smooth sidewall etched surface is achieved by NBE in a comparison of Figures 6.8a and 6.4e.

6.4   EVALUATION RESULTS

The resonance characteristics of the silicon micromechanical resonators fabricated by both DRIE and NBE are observed and evaluated, and the specifications are summarized in Table 6.2. The typical measurement results of the transmission S_{21} and phase response of the device fabricated by NBE are shown in Figure 6.9 for the resonator with a length of 500 μm, width of 440 μm, thickness of 5 μm, and capacitive gap width of 230 nm. A resonant peak, which is observed under measurement

TABLE 6.2

Summarized Parameters of the Silicon Resonators Fabricated by DRIE and NBE

	NBE (Structure 1)		DRIE (Structure 2)	
	Parameters			
Length of resonant body	$L=500$ μm		$L=500$ μm	
Width of resonant body	$W=440$ μm		$W=440$ μm	
Thickness of device layer	$t=5$ μm		$t=5$ μm	
Capacitive gap	$g=230$ nm		$g=400$ nm	
	5 V	10 V	5 V	10 V
	0 dBm	0 dBm	0 d Bm	0 dBm
	Theoretical Calculation			
Effective mass, m_{eff}	1.3×10^{-9} Kg	1.3×10^{-9} Kg	1.3×10^{-9} Kg	1.3×10^{-9} Kg
Effective stiffness, k_{eff}	4.7×10^{6} Nm^{-1}	4.7×10^{6}	4.7×10^{6} Nm^{-1}	4.7×10^{6} Nm^{-1}
Electromechanical	2.1×10^{-6}	Nm^{-1}	6.9×10^{-7}	1.4×10^{-6}
transduction, η	9.67 MHz	4.2×10^{-6}	9.67 MHz	9.67 MHz
Resonant frequency, f_0		9.67 MHz		
	Simulation Result			
Resonant frequency, f_0	9.75 MHz	9.75 MHz	9.75 MHz	9.75 MHz
	9.66 MHz	9.66 MHz	9.66 MHz	9.66 MHz
	77,000	75,000	None	63,000
	−72.2 dB	−63.8 dB	−92.6 dB	−78.7 dB
	230 kΩ	59 kΩ	None	645 kΩ
	0.9 aF	3.7 aF	None	0.4 aF
	293 H	73 H	None	670 H
	0.1 pF	0.1 pF	None	0.06 pF

conditions V_{DC} of 5 V and V_{AC} of 0 dBm, is found at 9.66 MHz. The FEM simulation result (9.75 MHz) of the resonant frequency is in good agreement with the experimental result (9.66 MHz) and theoretical calculation (9.67 MHz). Additionally, Figure 6.9 also presents a comparison of measured frequency characteristics in the vacuum chamber and ambient atmosphere. It shows the effects of viscous damping on the Q factor when the pressure is changed. The Q factor of 77,000 at a vacuum chamber of 0.01 Pa is larger than that of 28,000 in ambient atmosphere of 10^5 Pa.

At V_{DC} of 5 V, a resonant peak is clearly observed and a high Q factor of 77,000 is achieved for the device fabricated by NBE (structure 1) (Figure 6.10a), whereas that of the device fabricated by DRIE (structure 2) shows no vibration signal (Figure 6.10b). The capacitve gap width of structure 2 is around 400 nm, whereas that of structure 1 is around 230 nm. Therefore, structure 2 requires a higher V_{DC} than structure 1 in order to vibrate the resonator. The evaluation of the equivalent circuit model is performed for both devices as shown in Table 6.2. Based on the experimental data and the equivalent circuit model equations, structure 1 exhibits the motional resistance $R_m=230$ kΩ, motional capacitor $C_m=0.9$ aF, motional inductance $L_m=293$ H, and

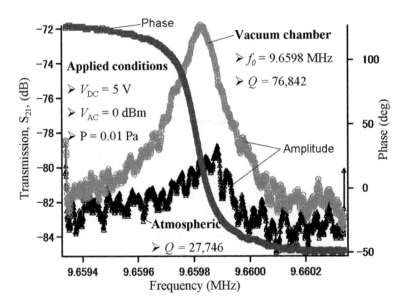

FIGURE 6.9 Transmission and phase response of NBE device and frequency response of NBE device both under vacuum chamber pressure and ambient atmospheric.

feed-through capacitance $C_f = 0.1$ pF, whereas those of structure 2 are unmeasurable at V_{DC} of 5 V. As the aforementioned measurement results at low V_{DC} are shown in Figure 6.10a and b, structure 1 has a higher perfomance than structure 2 because of the smaller capacitive gap widths. A quantitative comparison of these two devices is done at a higher V_{DC} of 10 V as shown in Table 6.2 and Figure 6.11. Although structures 1 and 2 have been designed in the same parameters of length L, width W, and thickness t, their resonant frequencies are a little different as we can see from Figure 6.11. This is possibly due to fabrication errors in the size, which make the effective mass and effective spring constant slightly changed. The measurement results show that the motional resistance of structure 1 is reduced by almost 11 times from 645 kΩ to 59 kΩ, and its insertion loss is increased by approximately 15 dB in comparison with those of structure 2. Additionally, the Q factor ($Q=75,000$) of structure 1 is higher than that of structure 2 ($Q=63,000$).

The measurement results shown in Figure 6.12 confirmed that the Q factor of the devices fabricated by NBE is higher than that of the devices fabricated by DRIE. Five devices fabricated by NBE and five devices fabricated by DRIE are evaluated and compared under the same measurement conditions V_{DC} of 10 V, V_{AC} of 0 dBm, and pressure vacuum of 0.01 Pa. The high Q factors of devices fabricated by NBE are obtained from 75,000 to 82,000 (average Q factor value of around 78,000), whereas those fabricated by DRIE are from 57,000 to 66,000 (average Q factor value of around 61,000), as shown in Figure 6.12. Generally, when the capacitive gap widths are reduced, the motional resistance value is decreased. This causes the drop in the Q factor [21–23]. However, the gap widths of structure 1 are smaller than those of structure 2; the higher Q factor of NBE is obtained. Thus, more defects on the surface

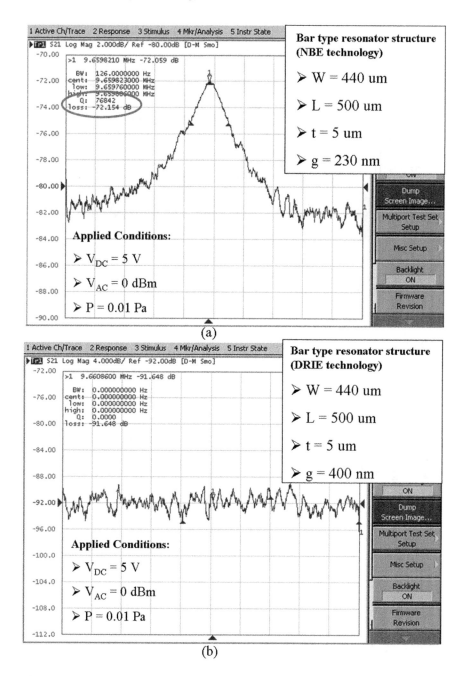

FIGURE 6.10 (a) Frequency response of NBE device at V_{DC} of 5 V. (b) Frequency response of DRIE device at V_{DC} of 5 V.

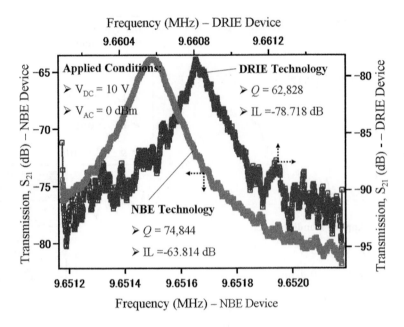

FIGURE 6.11   Frequency response of NBE and DRIE devices at V_{DC} of 10 V.

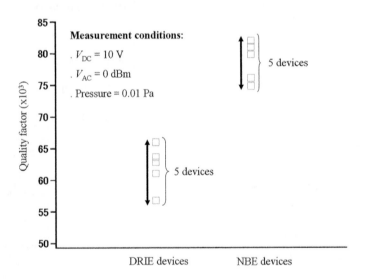

FIGURE 6.12   Comparison of a quality factor between NBE and DRIE devices.

are generated on devices fabricated by DRIE than those of NBE, which results in more energy dissipation that leads to mechanical deterioration. Therefore, the higher Q factor value can be achieved by using NBE technology. The possible reasons of the lower Q factor of devices fabricated by DRIE than that of those fabricated by NBE are due to the roughness and lattice defects on the surfaces. The etched surfaces of the devices fabricated by NBE are smooth, while those of the devices fabricated by DRIE are rough due to scallops. In addition, the plasmas irradiations (UV/VUV) of DRIE possibly cause defects on the surface of the device, although the surface of the device is protected by hard mask (SiO_2) during etching. UV/VUV photons with wavelengths ranging from about 10 nm to about 200 nm penetrate into the SiO_2 and Si layers and generate positive charges and/or interface traps in SiO_2/Si [24, 25]. These positive charges and interface traps may cause defects on the surface of the silicon device layer after removing SiO_2 by using hydrofluoric acid (HF) solution. However, no damages are generated when the neutral beam irradiation is employed due to high-aspect-ratio carbon aperture[11, 17, 26, 27]. Further investigation is needed to clarify the quantitative effect of NBE and DRIE on the surface damage of silicon resonators.

6.5 SUMMARY

In this chapter, frequency characteristics of the devices fabricated by NBE with the resonant frequency of 9.66 MHz with a length of 500 μm, width of 440 μm, and thickness of 5 μm are evaluated, and a high average Q factor value of around 78,000 is achieved. Additionally, the devices fabricated by both DRIE and NBE are evaluated and compared with each other. The devices fabricated by NBE show that the motional resistances are reduced by almost 11 times and their output signals are increased by approximately 15 dB than those fabricated by DRIE. Especially, devices fabricated by NBE provide higher Q factors (from 75,000 to 82,000) than those of devices fabricated by DRIE (from 57,000 to 66,000) in the comparison of the same resonator parameters and measurement conditions.

REFERENCES

1. van Beek, J.T.M., Puers, R., A review of MEMS oscillators for frequency reference and timing applications, *Journal of Micromechanics and Microengineering*, **22**, 013001, 2012.
2. Nguyen, C.T.C., MEMS technology for timing and frequency control, *IEEE Transactions on Ultrasonics, Ferroelectrics, and Frequency Control*, **54**, 251–270, 2007.
3. Ayazi, F., MEMS for integrated timing and spectral processing, *IEEE 2009 Custom Integrated Circuits Conference (CICC)*, San Jose, CA, 65–67, 2009.
4. Toan, N.V., Miyashita, H., Toda, M., Kawai, Y., Ono, T., Fabrication of an hermetically packaged silicon resonator on LTCC substrate, *Microsystem Technologies*, **19**, 1165–1175, 2012.
5. Lin, Y.W., Li, S.S., Xie, Y., Ren, Z., Nguyen, C.T.C., Vibrating micromechanical resonators with solid dielectric capacitive transducer gaps, *Frequency Control Symposium and Exposition*, 128–134, 2005.
6. Weinstein, D., Bhave, S.A., Piezoresistive sensing of a dielectrically actuated silicon bar resonator, *Solid-State Sensors, Actuators, and Microsystems Workshop*, 368–371, 2008.
7. Pourkamali, S., Ho, G.K., Ayazi, F., Low impedance VHF and UHF capacitive silicon bulk acoustic wave resonators – Part I: Concept and fabrication, *IEEE Transactions on Electron Devices*, **54**, 2017–2023, 2007.
8. Tomura, M., Huang, C.H., Yoshida, Y., Ono, T., Yamasaki, S., Samukawa, S., Plama induced deterioration of mechanical characteristic of microcantilever, *Japanese Journal of Applied Physics*, **49**, 04DL20, 2010.

9. Oehrlein, G.S., Tromp, R.M., Lee, Y.H., Petrillo, E.J., Study of silicon contamination and near surface damage cause by CF4/H2 reactive ion etching, *Applied Physics Letters*, **45**, 420–422, 1984.
10. Petti, C.J., Mcvitte, J.P., Plummer, J.D., Characterization of surface mobility on the sidewalls of dry etched trenches, *Electron Devices Meeting*, 104–107, 1988.
11. Wada, A., Yanagisawa, Y., Altansukh, B., Kubota, T., Ono, T., Yamasaki, S., Samukawa, S., Energy-loss mechanism of single crystal silicon microcantilever due to surface defects generated during plasma process, *Journal of Micromechanics and Microengineering*, **23**, 065020, 2013.
12. Samukawa, S., Novel neutral beam etching processes for future nanoscale devices, *The International Microprocesses and Nanotechnology Conference*, 472–473, 2007.
13. Samukawa, S., Sakamoto, K., Ichiki, K., Generating high-efficiency neutral beams by using negative ions in an inductively coupled plasma source, *Journal of Vacuum Science & Technology A*, **20**, 1566–1573, 2002.
14. Endo, K., Noda, S., Masahara, M., Kubota, T., Ozaki, T., Samukawa, S., Liu, Y., Ishii, K., Ishikawa, Y., Sugimata, E., Matsukawa, T., Takashima, H., Yamauchi, H., Suzuki, E., Damage free neutral beam etching technology for high mobility FinFETs, *Electron Devices Meeting*, 104–107, 2005.
15. Kubota, T., Nukaga, O., Ueki, S., Sugiyama, M., Inamoto, Y., Ohtake, H., Samukawa, S., 200 mm diameter neutral beam source based on inductively coupled plasma etcher and silicon etching, *Journal of Vacuum Science & Technology A*, **28**, 1169–1174, 2010.
16. Kubota, T., Watanabe, N., Ohtsuka, S., Iwasaki, T., Ono, K., Iriye, Y., Samukawa, S., Numerical study on electron transfer mechanism by collision of ions at graphite surface in highly efficient neutral beam generation, *Journal of Physics D: Applied Physics*, **45**, 095202, 2012.
17. Miwa, K., Nishimori, Y., Ueki, S., Sugiyama, M., Kubota, T., Samukawa, S., Low damage silicon etching using a neutral beam, *Journal of Vacuum Science & Technology B*, **31**, 051207, 2013.
18. Madou, M.J., *Fundamentals of Microfabrication: The Science of Miniaturization*, CRC Press, Boca Raton, FL, 2002.
19. A deep silicon RIE primer-Bosch etching of deep structures in silicon, 2009, http://www.nanofab.ualberta.ca/wp-content/uploads/2009/03/primer_deepsiliconrie.pdf.
20. Fitzgerald, A.M., MEMS structural reliability and DRIE: How are they related? 2010, http://www.memsjournal.com/tegalamfitzgeraldwebinar.html.
21. Toan, N.V., Toda, M., Kawai, Y., Ono, T., Capacitive silicon resonator with movable electrode structure for gap width reduction, *Journal of Micromechanics and Microengineering*, **24**, 025006, 2014.
22. Oka, T., Ishino, T., Tanigawa, H., Suzuki, K., Silicon beam micromechanical systems resonator with a sliding electrode, *Japanese Journal of Applied Physics*, **50**, 06GH02, 2011.
23. Shao, L.C., Palaniapan, M., Khine, L., Tan, W.W., Micromechanical resonators with submicron capacitive gaps in 2 um process, *Electronics Letters*, **43**, 1427–1428, 2007.
24. Samukawa, S., Ishikawa, Y., Kumagai, S., Okigawa, M., On-wafer monitoring of vacuum-ultraviolet radiation damage in high density plasma process, *Japanese Journal of Applied Physics*, **40**, L1346, 2001.
25. Yunogami, T., Mizutani, T., Suzuki, K., Nishimatsu, S., Radiation damage in SiO2/Si induced by VUV photons, *Japanese Journal of Applied Physics*, **28**, 2172, 1989.

26. Tomura, M., Huang, C.H., Yoshida, Y., Ono, T., Yamasaki, S., Samukawa, S., Plasma induced deterioration of mechanical characteristics of microcantilever, *Japanese Journal of Applied Physics*, **49**, 04DL20, 2010.
27. Nishimori, Y., Ueki, S., Miwa, K., Kubota, T., Sugiyama, M., Samukawa, S., Hashiguchi, G., Effect of neutral beam etching on mechanical property of microcantilevers, *Journal of Vacuum Science & Technology B*, **31**, 022001, 2010.

7 Capacitive Silicon Resonators with Narrow Gaps Formed by Metal-Assisted Chemical Etching

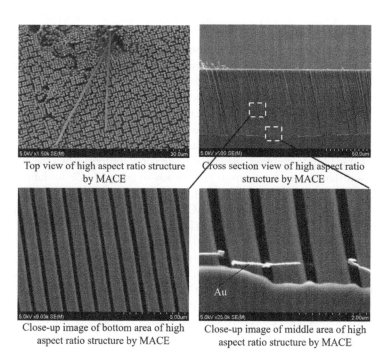

Top view of high aspect ratio structure by MACE	Cross section view of high aspect ratio structure by MACE
Close-up image of bottom area of high aspect ratio structure by MACE	Close-up image of middle area of high aspect ratio structure by MACE

7.1 INTRODUCTION

Silicon, one of the most important materials in micro/nano systems, has been used for the fabrication of a wide range of micro/nano devices, including microfabricated resonators [1, 2], power generators [3, 4], and bio/chemical sensors [5, 6]. Wet and dry etching techniques are typically employed for patterning silicon structures. The wet etching of silicon [7] is performed under liquid phase in a container, such as a beaker, consisting of potassium hydroxide (KOH) and tetramethylammonium hydroxide (TMAH). The wet etching of silicon can achieve smooth etched surfaces; however, the aspect-ratio structures are limited due to side etching. Whereas high-aspect-ratio structures can be easily achieved by inductively coupled plasma–reactive ion etching (ICP-RIE) with the Bosch process [8, 9] using SF_6 (etching cycle) and C_4F_8 (passivation cycle) gases. Nevertheless, many disadvantages [10] can be found such as limitation of achievable aspect structures, rough etched surfaces, and defect generation on silicon surfaces. Moreover, all these dry methods require radio frequency (RF) generators, vacuum chambers, and precise mechanical parts, such that the device's fabrication cost becomes high. Thus, novel methods for high-aspect-ratio silicon structures are still needed for micro/nano systems.

Many research groups [11–26] are working on metal-assisted chemical etching (MACE) technology owing to its simplicity, low fabrication costs, and ability to generate high-aspect-ratio nanostructures. The MACE process is held in a wet etching solution (mixture solution of HF and H_2O_2), but it enables the formation of anisotropic silicon structures at room temperature and atmospheric pressure. However, MACE was mostly used to prepare silicon nanostructures without demonstrating the specific applications, [14–18, 19–23]. The overview concerning the applications of silicon nanostructures by MACE in the field of energy conversion, storage, and sensors has been reported [11–13]. Applications include silicon nanowire fabrication by MACE [14–18] and the [19–23] the MACE mixture for vertical etching applied to the fabrication of high-aspect-ratio nanowires. Patterning large size structures by MACE still face challenges [11–26] because hydrofluoric acid (HF) liquid cannot reach center areas. Even though the oxide layer in the center area can be removed, hydrogen gas is still difficult to release, which may make the catalyst metal layer float or buckle [24]. In addition, the etching rate is not uniform when the etching of patterns of different sizes is required. Thus, a way of using MACE for fabrication of micro/nano devices needs to be addressed.

Our recent achievements related to MACE technology show that MACE has many advantages over deep reactive ion etching (DRIE) for forming high-aspect-ratio structures [25]. Ultra-high aspect trenches and pillars of 400 and 80, respectively, have been achieved using MACE, whereas those of DRIE are limited to around 30. High density as well as high-aspect-ratio silicon nanopillars have been demonstrated [26]. Ion transport by gating voltage to silicon nanopores (15 nm diameter and 200 µm height) produced via MACE was reported by Toan et al. [27]. Herein, a simple way of using MACE to produce a capacitive silicon resonator is proposed. Not only can large areas be patterned but also a combination of large and narrow areas can be formed. The chapter is organized as follows: In Section 7.2, the fundamentals of MACE are briefly introduced as well as experimental investigations that have studied MACE. In Section 7.3, fabrication of a capacitive silicon resonator structure is demonstrated, and its performance evaluated. Finally, we summarize the topic in Section 7.4.

7.2   METAL-ASSISTED CHEMICAL ETCHING

7.2.1   THEORY OF METAL-ASSISTED CHEMICAL ETCHING

MACE basically consists of two steps. First, a noble metal, such as Au, Ag, or Pt, which acts as a catalyst, is patterned on a silicon substrate. The sample is then immersed in an etching solution of HF and a H_2O_2 oxidant. Details of the MACE process are explained in the following.

The cathode reaction at the noble metal and liquid interface generates electrical holes h^+ as in Equation 7.1. The reaction creates H_2O and at the same time injects holes h^+ into the Si through the noble metal layer. The silicon atoms underneath the noble metal layer are oxidized by the injected h^+.

$$H_2O_2 + 2H^+ \rightarrow 2H_2O + 2h^+, \tag{7.1}$$

$$Si + 6HF + 2h^+ \rightarrow H_2SiF_6 + 2H^+ + H_2, \tag{7.2}$$

The anode reaction at the silicon–metal interface dissolves the silicon underneath the noble metal pattern with the HF solution. H^+ and hydrogen gas H_2 are produced by this reaction (Equation 7.2). The noble metal patterns penetrate into the space formed by the etched silicon. Equations 7.1 and 7.2 can be combined into

$$Si + 6HF + H_2O_2 \rightarrow H_2SiF_6 + 2H_2O + H_2. \tag{7.3}$$

The preceding mechanism is named mass transportation, which is presented in many works [11, 12]. However, another mechanism is also considered, namely, charge transportation [13]. It can be summarized as follows: A cathode, at the interface between the metal and etching solution, would catalyze H_2O_2 reduction, consuming H^+ and electrons in the process (Equation 7.4). The potential of the cathode site is to be closed to $EH_2O_2 = 1.77$ V (standard hydrogen electrode).

$$H_2O_2 + 2H^+ + 2e^- \rightarrow 2H_2O, \tag{7.4}$$

An anode, at the interface between metal and silicon, is the catalyst for silicon oxidation, which creates H+ and electrons (Equation 7.5). The potential of the anode site is to be closed to $E_{Si} = 1.2$ V.

$$Si + 6HF \rightarrow H_2SiF_6 + 4H^+ + 4e^-, \tag{7.5}$$

Thus, a local current flows from the cathode site to the anode site. The generated field from the differences in the charge transportation during the etching process causes the deviation of the catalyst from its vertical downward etching direction. In summary, the autonomous motion of the metal layer inside the silicon, which primarily originates from self-electrophoresis of the metal film, results in an anisotropic etching of silicon.

7.2.2   Survey of Metal-Assisted Chemical Etching

The details of the MACE process are shown in Figure 7.1. It starts with a p-type (100)–oriented silicon wafer with a thickness of 200 μm and a resistivity of 1–2 Ωcm (Figure 7.1a). After a conventional cleaning process, such as RCA-1, RCA-2, or piranha, samples were then immersed in a diluted HF solution to completely remove the

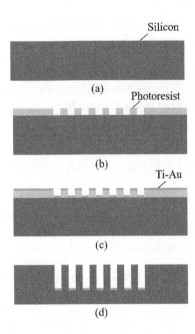

FIGURE 7.1 MACE process: (a) silicon wafer, (b) photolithography, (c) EB evaporation, (d) MACE.

SiO_2 from the surface. A 1 μm-thick positive photoreisist (TSMR V90, Tokyo Ohka Kogyo, Kanagawa, Japan) was applied to the surface of the wafer and patterned via lithography with a small amount of deionized water (Figure 7.1b). Pillars of 1 μm diameter and spaces of 1 μm between the pillars were achieved. Next, Ti and Au layers with thicknesses of 1 nm and 30 nm, respectively, were deposited on the resist pattern using electron beam evaporation at a base pressure of ~0.01 Pa with deposition rates at 0.01 nm/sec for Ti and 0.1 nm/sec for Au (Figure 7.1c). After deposition of the Ti and Au layers, the photoresist was removed with a resist stripper (MS-2001). The samples were then immersed in a mixture solution of HF (50%), H_2O_2 (30%), and ethanol with a volume ratio of 5:1:1 at room temperature. The noble metal Au pattern went down and etched the Si as shown in Figure 7.1d.

During the photolithography step, a small amount of deionized water was positioned on the mask and wafer to enhance the resolution of the lithographed patterns. Therefore, smaller and finer photoresist patterns can be achieved over those of conventional photolithography due to the improved refractive index matching between the mask and wafer. Moreover, vertical resist patterns can also be obtained using this technique. It makes the subsequent lift-off process easier. A thin Ti film was employed to improve the adherence of the Au catalyst film onto the silicon substrate. It was quickly etched out by the HF solution upon immersion during the etching process; therefore, the Au film can directly contact the Si surface and act as an active catalyst to etch the Si underneath. Ethanol was employed to reduce the surface tension between the etching solution and the Au surface. Therefore, a high etching rate and smooth etched surfaces could be achieved.

The etching results from MACE with the aforementioned conditions are shown in Figure 7.2. Top and 30°-titled views of scanning electron microscope (SEM) images of the silicon pillar structures fabricated by MACE are shown in Figure 7.2a and b, respectively. A high silicon etching rate (approximately 500 nm/min) and 2.5 μm etching depth have been observed when using the aforementioned etching solution with the process time of 5 min. A smooth and vertical etching profile was achieved (Figure 7.2c and d). The survey of the effects of the etching time, pattern size, and concentration of etching solution during MACE is presented as follows.

7.2.2.1 Effect of Etching Time

To investigate the etching rate dependence over etching time, various durations of etching time from a few minutes to several tens of minutes using the aforementioned MACE mixture (HF:H$_2$O$_2$:Ethanol = 5:1:1) to produce 1 μm-diameter pillars and 1 μm-spaces between pillars were used. The etching depth reached 12 μm after 15 min of the MACE process. The estimated etching rate was around 0.8 μm/min. By increasing the etching time to 30 min, a 50 μm etching depth was observed. The etching rate increased almost two times by increasing the etching duration from 15 min to 30 min. The MACE results for 15 min and 30 min are shown in Figure 7.3a and b, respectively. The etching depth was evaluated as a function of etching time,

FIGURE 7.2 MACE result for 5 min etching: (a) top view, (b) 30°-titled view, (c) close-up view, (d) cross-section view.

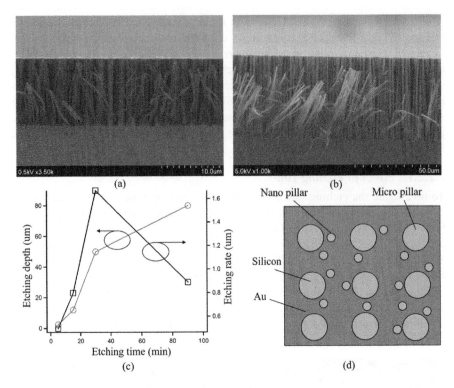

FIGURE 7.3 Effect of the etching time: (a) 15 min etching, (b) 30 min etching. (c) Etching depth as a function of etching time. (d) Micro and nano pillar model.

as shown in Figure 7.3c. An etching depth of around 80 µm was achieved after 90 min. The etching rate was around 0.88 µm/min. Therefore, the etching rate of the silicon first increased as a linear trench and then slightly decreased during the long-time MACE process, as shown in Figure 7.3c. A possible reason for this phenomenon could be the production of H_2O during the MACE process. H_2O lowers the concentration of the etching solution. Thus, the etching rate would be lower over the longer-time MACE process.

There are many silicon nanopillars observed after the MACE process, as shown in Figure 7.3a and b. This may be due to various reasons: First, there was no Au metal in these areas due to the nonuniform deposition of the thin Au layer during the EB evaporation process. Second, the nano areas of the Au film were peeled off during the lift-off process because of a weak adherence between the Au and the silicon or particles. Last, the nano areas of the Au film may be damaged, which could be caused by the MACE process due to the penetration of the electrical holes h^+ through the metal. Therefore, the Au patterns on the silicon substrate consisted of not only micropillars but also nanopillars, as demonstrated in Figure 7.3d. This would result in the silicon micro- and nanopillars after the MACE process, as shown in Figure 7.3a and b.

7.2.2.2   Effect of Pattern Size

A sample with a square etching area of 500 μm² was used for the survey of the effect of the pattern size. After 30 min etching, the etched silicon was observed. However, some areas were not etched and the Au film peeled off. The sample model is shown in Figure 7.4a and the etched result is given in Figure 7.4a'. H_2 gas was generated by the anode reaction, as shown in Equation 7.2, which was generated underneath the Au film during the MACE process. If the etching area were too large, the gas within the square Au pattern could not escape. It therefore caused the Au film to peel off and the etching process was not successful. A smaller pattern of 10 μm pillars and 10 μm spacers was created with the same MACE conditions for a 500 μm² area, as earlier. The etching results are indicated in Figure 7.4b and b'. The silicon could be etched, but the silicon pillar structures were destroyed and a porous structure was created, as shown in Figure 7.4b'. The initial etching depth was good, and then the patterns were destroyed. This means that the continuous Au film was destroyed and became Au particles. It was then penetrated in certain directions during the MACE with Au particles, which made the silicon become porous. The MACE process for 1 μm pillars and 1 μm spacers was done with a high etching rate, vertical etched profile, and smooth etched surfaces, as earlier. Thus, small etched pattern sizes were easier to produce than large pattern sizes by the MACE method. To overcome the etching problem for the large area using MACE, the effects of the concentration of the etching solution have to be investigated.

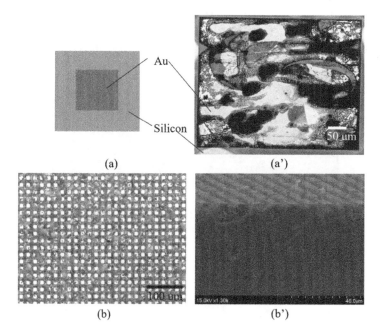

FIGURE 7.4   Effect of pattern sizes: (a) 500 μm² model, (a') 500 μm² pattern size after etching, (b) top view of 10 μm² pattern, (b') cross-section view of 10 μm² pattern.

7.2.2.3   Effect of Concentration of Etching Solution

To study the effect of the concentration of the etching solution on the etching, the volume of the etching solution was kept constant and its composition was varied via the volumetric ratio of HF, H_2O_2, and ethanol. Based on the first MACE recipe (HF:H_2O_2:Ethanol = 5:1:1) for the aforementioned small pattern, the optimum MACE recipe for the large pattern could be found. In the first experiment, the amount of HF was decreased and that of ethanol was increased. The etching results with the new recipe of HF:H_2O_2:Ethanol = 3:1:3 (recipe 2) are shown in Figure 7.5a and a'. Fine patterns can be observed and the Au film was not damaged. The etching results for the large pattern using recipe 2 were better than when using recipe 1. One of the possible reasons is the greater amount of ethanol in recipe 2 over that of recipe 1, which helps the H_2 to more easily escape. In the second experiment, the recipe of HF:H_2O_2:Ethanol = 1:3:3 (recipe 3) was used and the etching results are shown in Figure 7.5b and b'. Some damaged areas on the etched surfaces could be observed. The rate of the electrical hole consumption at the silicon–Au film interface was smaller than the rate of the electrical hole injection. Then, the electrical holes were released and moved up, which resulted in damaged surfaces. Forming large patterns by MACE was therefore possible with the proper recipe for the concentration of the etching solution. The summarized etching conditions are shown in Table 7.1.

The effect of the concentration of the etching solution on a small pattern was also investigated. A larger amount of HF than the other liquids in recipe 1 (HF:H_2O_2:Ethanol = 5:1:1) was used. Here, the proportions of the recipe have been altered. First, the volumes of HF and H_2O_2 have been altered to be the same, which gives the recipe of HF:H_2O_2:Ethanol = 3:3:1 (recipe 4). The etched profiles are shown in Figures 7.6 a and a'. After the etching time of 30 min, the etched silicon depth

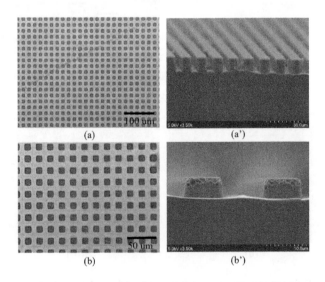

(a)   (a')

(b)   (b')

FIGURE 7.5 Optimum MACE conditions: (a) top view of 10 µm² pattern with recipe 2, (a') Cross section view of 10 µm² pattern with recipe 2, (b) top view of 10 µm² pattern with recipe 3, (b') cross-section view of 10 µm² pattern with recipe 3.

TABLE 7.1
Summary of Etching Conditions

	HF:H$_2$O$_2$:Ethanol	Etching Depth	Comments
Large pattern size (10 μm size)	5:1:1 (etching time of 30 min)	31 μm	Initial etching depth is good, and then the patterns are destroyed.
	3:1:3 (etching time of 30 min)	10 μm	Vertical and smooth etched surfaces.
	1:3:3 (etching time of 30 min)	5 μm	Etched surfaces are damaged.
Small pattern size (1 μm size)	5:1:1 (etching time of 30 min)	50 μm	Pillar structures are fine, but nanopillars occur.
	3:3:1 (etching time of 30 min)	63 μm	Pillar diameter becomes smaller than expected.
	1:5:1 (etching time of 30 min)	130 μm	Pillars are destroyed and structures become porous.

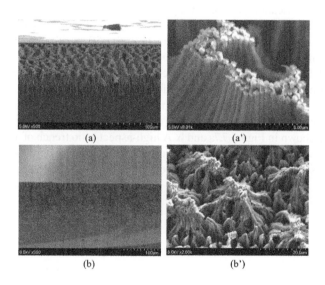

(a)	(a')
(b)	(b')

FIGURE 7.6 Effect of concentration of etching solution: (a) tilted view of 1 μm^2 pattern with recipe 4, (a') close-up view of 1 μm^2 pattern with recipe 4, (b) cross-section view of 1 μm^2 pattern with recipe 5, (b') tilted-view of 1 μm^2 pattern with recipe 5.

reached more than 60 μm. The diameters of the silicon pillars formed by recipe 4 became smaller than those from recipe 1. The diameter of the silicon pillars decreased from 1 μm to 500 nm when moving from recipe 1 to recipe 4. Next, the amount of H$_2$O$_2$ over that of HF was increased with recipe 5 of HF:H$_2$O$_2$:Ethanol = 1:5:1, and the etching results after 30 min are shown in Figure 7.6b and b'. Thus, a higher etching depth can be observed when increasing the amount of H$_2$O$_2$, but the pattern has been destroyed. This is possibly due to recipes 4 and 5 leaving too many electrical holes h$^+$ and the low rate of hole consumption. Therefore, the structures are damaged in a similar way to that for the large pattern sizes. The summarized etching conditions are shown in Table 7.1.

7.2.2.4 High-Aspect-Ratio Silicon Structures

In this section, two methods, DRIE and MACE, were investigated and compared for high-aspect-ratio silicon structures. The details of the comparison between the two methods for high-aspect-ratio structures are summarized in Table 7.2. High-aspect-ratio structures have been widely applied in antireflection structures, through-silicon vias (TSVs), for modifying the wetting behavior of the bulk substrate and silicon capacitive resonators. High-aspect-ratio capacitive gaps are one of the most important factors to achieve high performance in this kind of resonator.

Generally, Si high-aspect-ratio structures are fabricated using DRIE with the Bosch process. This is a series of sandwiching processes for the etching and passivation cycles. Passivation steps are employed to protect the sidewall so that the anisotropy of the etched silicon can be achieved. This technique is easily reproducible and can be used to efficiently fabricate a large number of similar devices (mass production). The fabrication process for high-aspect-ratio Si trenches formed by DRIE was as follows: A 1 μm-thick SiO_2 film was grown on the silicon substrate by a wet thermal oxidation process for the etching mask of the silicon. To make nanotrenches, electron beam lithography (EBL) was performed. Then, the SiO_2 mask layer was etched by RIE. Next, DRIE with the Bosch process of short etching and passivation cycles (etching time 2.5 sec; passivation time 2.5 sec) was performed to achieve low scallops and vertical etching profiles. Thus, very small scallops of around 10 nm can be observed and a vertical sidewall with a small gap of around 350 nm was obtained, as shown in Figure 7.7a and b. A high-aspect-ratio trench of around 20 was achieved. It can reach a slightly higher value by increasing the etching time of the DRIE process, but it would be limited due to the erosion of the mask. The aspect

---

TABLE 7.2
Summary of DRIE and MACE for High-Aspect-Ratio Structures

	DRIE	MACE
Etching environment	Dry etching method	Wet etching method
	Plasma	Chemical reaction
Feature size	Pattern size: 350 nm	Pattern size: 200 nm
	Etching depth: 5 μm	Etching depth: 80 μm
Mask selectivity	Limited	Not limited
Etching profile	Anisotropic etching	Anisotropic etching
	Vertical sidewall	Vertical sidewall
	Existing scallops	Smooth
Aspect ratio structure	Limited	Not limited
Fabrication cost	High	Low
Reproducible process	Very high	Low
Mass production	Available	Not yet

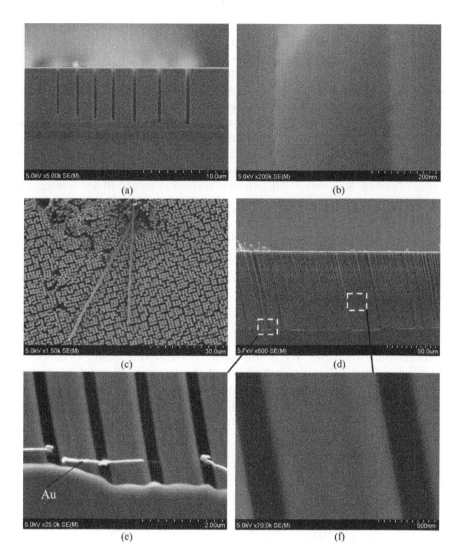

FIGURE 7.7 High-aspect-ratio structure. (a) Cross section of high-aspect-ratio structure by DRIE. (b) Close-up view of high aspect ratio structure by DRIE. (c) Top view of high-aspect-ratio structure by MACE. (d) Cross-section view of high-aspect-ratio structure by MACE. (e) Close-up image of bottom area of high-aspect-ratio structure by MACE. (f) Close-up image of middle area of high aspect ratio structure by MACE.

ratio and achievable etching depth were therefore limited by the anisotropy and etch selectivity.

In turn, in MACE, the lithography with a small amount of deionized water was replaced by EBL for creating a smaller pattern. Pillars of 1 μm diameter and spaces between the pillars of 200 nm in the EB resist pattern have been achieved. The other steps in the sample preparation process were similar to the MACE process discussed earlier. The sample was immersed in the etching solution for 150 min, from which the etching results are shown in Figure 7.7c and d. Pillars with 80 μm etching depth were achieved with smooth etched surfaces and vertical etching profile. Thus, very high-aspect-ratio trench (400, open trench of 200 nm, etching depth of 80 μm) and pillar (80, diameter of 1 μm, pillar height of 80 μm) structures were achieved. The Au metal pattern goes down, as clearly shown in Figure 7.7e. It was positioned slightly above the bottom surface due to a low adhesion between the Au and the silicon. When drying or cutting the sample, it was easily peeled off from the surface. Consequently, the structures fabricated by MACE had a much higher aspect ratio than those fabricated by DRIE in the comparison of the etching results between Figure 7.7a and d. Moreover, DRIE exhibited scallops of around 10 nm, whereas the MACE had a smooth surface.

7.3   CAPACITIVE SILICON RESONATOR

Resonator structure in this work is the two-arm-type capacitive silicon resonator, as shown in Figure 7.8. It basically consists of a silicon resonant body, driving and sensing electrodes, capacitive gaps, and supporting beams. The resonant body is separated with driving and sensing electrodes by a narrow capacitive gap. When an alternating current (AC) voltage V_{AC} is applied to the driving electrode, the resulting electrostatic force induces a bulk acoustic wave in the resonator element. Additional direct current (DC) voltage V_{DC} is also applied to the driving/sensing electrodes in order to amplify the electrostatic force. Small changes in the size of the capacitive gap generate a voltage on the sensing electrode.

Figure 7.8a shows the design fabrication of a two-arm-type capacitive silicon resonator. The large etching areas consisted of metal mesh patterning. The narrow

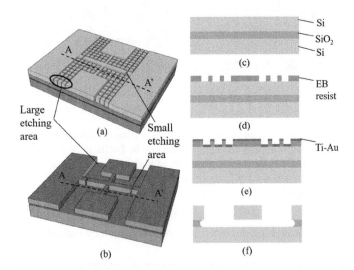

FIGURE 7.8   Capacitive silicon resonator. (a) Design fabrication. (b) Expected capacitive silicon resonator structures. (c) SOI wafer (7 μm/1 μm/300 μm). (d) Electron beam lithography. (e) Electron beam evaporator. (f) MACE process and cleaning process.

etching gap is the metal line by which both sides are connected to large etching areas. In this situation, when large areas are etched, a narrow capacitive gap will form. The expected structure is shown in Figure 7.8b.

A silicon on insulator (SOI) wafer, which consists of a 7 μm-thick top silicon device layer, 1 μm-thick oxide layer, and 300 μm-thick silicon handling layer, is used for the resonator fabrication (Figure 7.8c). An electron beam resist (ZEP 520A) layer is coated and patterned on the silicon device layer (Figure 7.8d). Ti–Au metal catalyst layers are subsequently deposited by the EB evaporator (Figure 7.8e). Next, silicon is etched by MACE with the same conditions as in the previous section (Figure 7.8f). A long etching time is employed to ensure that the oxide layer underneath the resonator body structure is completely removed. After etching, the wafer is cleaned by piranha solution (H_2SO_4:H_2O_2 = 2:1 volume ratio) to remove resist and metal. Silicon nanopillars in the etching areas could be easily removed by ultrasonic agitation with power of 300 W, frequency of 750 KHz, and isopropyl alcohol (IPA) liquid.

The capacitive nanogaps together with the resonator structure have been successfully fabricated via MACE, as demonstrated in Figure 7.9a. This resonator is excited in the longitudinal–extensional mode at the fundamental resonant frequency, as shown in FEM simulation in Figure 7.9d. Nanogaps of 250 nm wide and 7 μm high are achieved, as demonstrated in Figure 7.9e and f. The frequency responses of the fabricated device is evaluated by electrical setup shown in Figure 7.10a, which consisted of a network analyzer (E5071B ENA Series, Agilent Technologies) with a frequency range from 300 kHz to 8.5 GHz, DC voltage source V_{DC}, and electrical components including capacitors and resistors. The resonators are set in a vacuum

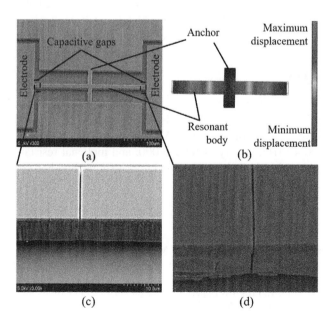

FIGURE 7.9   (a) Fabricated result of two-arm-type resonator structure. (b) FEM simulation. (c) and (d) Close-up image of nano-capacitive gap area.

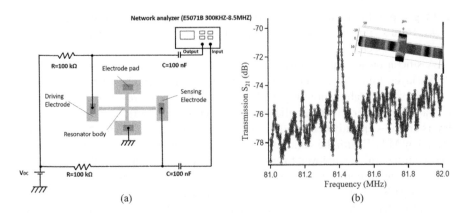

FIGURE 7.10 (a) Measurement setup. (b) Frequency response.

chamber with a pressure of 0.01 Pa. Frequency responses of the fabricated devices are shown in Figure 7.10b. The resonant peak is found at 81.4 MHz with a quality factor of 4000. However, the predicted simulation result is just half (39.4 MHz) of the experimental result. By looking carefully the etching trench at the bottom area of Figure 7.9d, a thin silicon layer still remains at bottom area, which again the vibration of a resonator structure at one side resulted in a much high spring constant of the resonator. In this case, if one arm is fixed, the resonant frequency is increased to 88.9 MHz according to FEM simulation (Figure 7.10b), which is close to the resonator vibration. Motional resistance of the fabricated resonator, based on insertion loss as demonstrated by Toan [28–30], is 89 KΩ, which is much smaller in comparison to other results [8, 31, 32] (same two-arm type resonator structure).

7.4 SUMMARY

MACE was investigated in this chapter for the etching of patterns of different sizes. Not only were large patterning areas but also a combination of large and narrow patterning areas were successfully formed. A capacitive silicon resonator with a resonant peak of 81.4 MHz, quality factor of 4000, and motional resistance of 89 KΩ was fabricated by MACE. This proposal exhibits a simple and low-cost fabrication of micro/nano devices.

REFERENCES

1. van Beek, J.T.M., Puers, R., A review of MEMS oscillators for frequency reference and timing applications, *Journal of Micromechanics and Microengineering*, **22**, 013001, 2012.
2. Toan, N.V., Ono, T., Progress in performance enhancement methods for capacitive silicon resonators, *Japanese Journal of Applied Physics*, **56**, 110101, 2017.
3. Pennelli, G., Macucci, M., High-power thermoelectric generators based on nanostructured silicon, *Semiconductor Science and Technology*, **31**, 054001, 2016.

4. Ziouche, K., Yuan, Z., Lejeune, P., Lasri, T., Leclercq, D., Bougrioua, Z., Silicon based monolithic planar micro thermoelectric generator using bonding technology, *Journal of Microelectromechanical Systems*, **26**, 45–47, 2017.
5. Presnova, G., Presnov, D., Krupenin, V., Grigorenko, V., Trifonov, A., Andreeva, I., Ignatenko, O., Egorov, A., Rubtsova, M., Biosensor based on a silicon nanowire field-effect transistor functionalized by gold nanoparticles for highly sensitive determination of prostate specific antigen, *Biosensor and Bioelectronics*, **88**, 293–289, 2017.
6. Alikhani, A., Gharooni, M., Abiri, H., Farokhmanesh, F., Abdolahad, M., Tracing the pH dependent activation of autophagy in cancer cells by silicon nanowire-based impedance biosensor, *Journal of Pharmaceutical and Biomedical Analysis*, **154**, 158–165, 2018.
7. Gosalvez, M.A., Zubel, I., Vinikka, E., Chapter 22 – Wet etching of silicon, in: *Handbook of Silicon Based MEMS Materials and Technologies*, 2nd ed., edited by M. Tilli, T. Motooka, V.-M. Airaksinen, S. Franssila, M. Paulasto-Krockel, V. Lindroos, pp. 470–502, Elsevier, 2015, https://doi.org/10.1016/B978-0-323-29965-7.00022-1.
8. Toan, N.V., Miyashita, H., Toda, M., Kawai, Y., Ono, T., Fabrication of an hermetically packaged silicon resonator on LTCC substrate, *Microsystem Technologies*, **19**, 1165–1175, 2013.
9. Toan, N.V., Toda, M., Kawai, Y., Ono, T., Capacitive silicon resonator with movable electrode structure for gap width reduction, *Journal of Micromechanics and Microengineering*, **24**, 025006, 2014.
10. Toan, N.V., Kubota, T., Sekhar, H., Samukawa, S., Ono, T., Mechanical quality factor enhancement in silicon micromechanical resonator by low-damage process using neutral beam etching technology, *Journal of Micromechanics and Microengineering*, **24**, 085005, 2014.
11. Han, H., Huang, Z., Lee, W., Metal-assisted chemical etching of silicon and nanotechnology applications, *Nano Today*, **9**, 271–304, 2014.
12. Li, X., Metal assisted chemical etching for high aspect ratio nanostructures: A review of characteristics and applications in photovoltaics, *Current Opinion in Solid State and Materials Science*, **16**, 71–81, 2012.
13. Huang, Z., Geyer, N., Werner, P., de Boor, J., Gösele, U., Metal-assisted chemical etching of silicon: A review, *Advanced Materials*, **23**, 285–308, 2011.
14. Fang, H., Wu, Y., Zhao, J., Zhu, J., Silver catalysis in the fabrication of silicon nanowire arrays, *Nanotechnology*, **17**, 3768–3774, 2006.
15. Williams, M.O., Jervell, A.L.H., Hiller, D., Zacharias, M., Using HCL to control silver dissolution in metal-assisted chemical etching of silicon, *Physica Status Solidi A*, **215**, 1800135, 2018.
16. Lin, H., Fang, M., Cheung, H.Y., Xiu, F., Yip, S., Wong, C.Y., Ho, J.C., Hierarchical silicon nanostructured arrays via metal-assisted chemical etching, *RSC Advances*, **4**, 50081–50085, 2014.
17. Williams, M.O., Hiller, D., Bergfeldt, T., Zacharias, M., How to oxidation stability of metal catalysts defines the metal assisted chemical etching of silicon, *Journal of Physical Chemistry C*, **121**, 9296–9299, 2017.
18. Miao, B., Zhang, J., Ding, X., Wu, D., Wu, Y., Lu, W., Li, J., Improved metal assisted chemical etching method for uniform, vertical and deep silicon structure, *Journal of Micromechanics and Microengineering*, **27**, 055019, 2017.
19. Um, H.D., Kim, N., Lee, K., Hwang, I., Seo, J.H., Yu, Y.J., Duane, P., Wober, M., Seo, K., Vertical control of metal-assisted chemical etching for vertical silicon microwire arrays and their photovoltaic applications, *Scientific Reports*, **5**, 11277, 2015.
20. Lianto, P., Yu, S., Wu, J., Thompson, C.V., Choi, W.K., Vertical etching with isolated catalysts in metal assisted chemical etching of silicon, *Nanoscale*, **4**, 7532–7539, 2012.

21. Chang, C., Sakdinawat, A., Ultra-high aspect ratio high-resolution nanofabrication for hard X-ray diffractive optics, *Nature Communications*, **5**, 4243, 2014.
22. Romano, L., Kagias, M., Jefimovs, K., Stampanoni, M., Self-assembly nanostructured gold for high aspect ratio silicon microstructures by metal assisted chemical etching, *RSC Advances*, **6**, 16025–16029, 2016.
23. Akan, R., Parfeniukas, K., Vogt, C., Toprak, M.S., Vogt, U., Reaction control of metal-assisted chemical etching for silicon-based zone plate nanostructures, *RSC Advances*, **8**, 12628–12634, 2018.
24. Yamada, K., Yamada, M., Maki, H., Itoh, K.M., Fabrication of arrays of tapered silicon micro-/nano-pillars by metal-assisted chemical etching and anisotropic wet etching, *Nanotechnology*, **29**, 28LT01, 2018.
25. Toan, N.V., Toda, M., Ono, T., High aspect ratio silicon structures produced via metal assisted chemical etching and assembly technology for cantilever fabrication, *IEEE Transactions on Nanotechnology*, **16**, 567–573, 2017.
26. Toan, N.V., Toda, M., Hokama, T., Ono, T., Cantilever with high aspect ratio nano pillars on its surface for moisture detection in electronic products, *Advanced Engineering Materials*, **19**, 1700203, 2017.
27. Toan, N.V., Inomata, N., Toda, M., Ono, T., Ion transportation by gating voltage to nanopores produced via metal assisted chemical etching method, *Nanotechnology*, **29**, 195301, 2018.
28. Toan, N.V., Shimazaki, T., Ono, T., Single and mechanically coupled capacitive silicon nanomechanical resonators, *Micro and Nano Letters*, **11**, 591–594, 2016.
29. Toan, N.V., Nha, N.V., Song, Y., Ono, T., Fabrication and evaluation of capacitive silicon resonator with piezoresisitve heat engines, *Sensor and Actuator A: Physical*, **262**, 99–107, 2017.
30. Toan, N.V., Shimazaki, T., Inomata, N., Song, Y., Ono, T., Design and fabrication of capacitive silicon nanomechanical resonators with selective vibration of a high-order mode, *Micromachines*, **8**, 312, 2017.
31. Mattila, T., Kiihamäki, J., Lamminmäki, T., Jaakkola, O., Rantakari, P., Oja, A., Seppä, H., Kattelus, H., Tittonen, I., A 12 MHz micromechanical bulk acoustic mode oscillator, *Sensors and Actuators A: Physical*, **101**, 1–9, 2002.
32. Kaajakari, V., Mattila, T., Lipsanen, A., Oja, A., Nonlinear mechanical effects in silicon longitudinal mode beam resonators, *Sensor and Actuator A: Physical*, **120**, 64–70, 2004.

Part 2B

Performance Enhancement Methods for Capacitive Silicon Resonators

Design Considerations

8 Mechanically Coupled Capacitive Nanomechanical Silicon Resonators

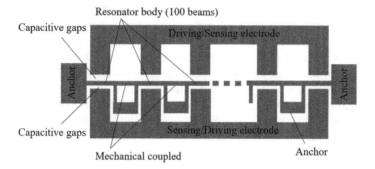

Mechanical coupled capacitive silicon nanomechanical resonator

Eigen frequency = 6.71 MHz

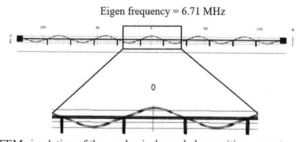

FEM simulation of the mechanical coupled capacitive resonator

8.1 INTRODUCTION

Silicon micro/nanomechanical resonators have been studied for a broad range of applications [1–7] such as timing devices [1–3] in oscillation circuits for modern data and communication applications, high-sensitivity sensor [4–6], and quantum information processing [7]. Major advantages are their small mass, high frequency, low intrinsic dissipation, and feasibility of integration with integrated circuit (IC) technology. One of the simplest and most common methods to detect the motion of microfabricated resonators is based on the measurement of the change in capacitance between a sensing electrode and the resonant body [8–10]. However, this method may not be effective for nanomechanical resonators because of their very small change of motional capacitance. Additionally, the nanomechanical resonator has a large motional resistance, and hence, high insertion loss that makes it difficult for practical applications. Generally, the motional resistance should be as small as possible. Therefore, a method for attaining a low motional resistance would be highly desirable.

In order to achieve a low motional resistance, investigations should be trying to improve the coupling efficiency. Narrowing the capacitive gap is by far the most effective, with a fourth power dependence [1, 11]. Different approaches to realize narrow capacitive gaps in silicon resonators have been investigated [12–14]. Resonator structures with a solid-filled gap are presented in Lin et al. [12]. Toan et al. reported capacitive gap reduction by electrostatic force [13]. A fabrication method using a thin oxide film as a sacrificial layer has been presented by Pourkamali et al. [14]. Besides efforts to reduce the capacitive gap size, development of arrays of the nanomechanical resonator is also a solution for low motional resistance. Arrays of 20, 49, and 100 resonators were designed and fabricated [15], then all individual devices were electrically interconnected. Low motional resistances have been achieved; however, they are challenged with a large parasitic capacitance and the mismatch of the resonant frequencies that result in difficulty in capacitive detection.

In this chapter, mechanically coupled capacitive silicon nanomechanical resonators with 100 individual beams are fabricated and evaluated. The clamped–clamped beam resonators are mechanically connected to a neighboring resonator by coupling silicon beams. The mechanical resonance properties and synchronization performance are evaluated.

8.2   DEVICE STRUCTURE AND WORKING PRINCIPLE

The single beam nanomechanical resonator (device 1) is shown in Figure 8.1a. It basically consists of silicon resonant body, driving/sensing electrodes, and capacitive gaps. The resonator is electrically excited and vibrated at a flexural mode by the combined influence of direct current (DC) and alternating current (AC) actuation voltages (V_{DC} and V_{AC}). The output voltage of the resonator results from the changes in the capacitive gap on the sensing electrode. In turn, the designed mechanically coupled nanomechanical resonator (device 2) consists of 100 single nanomechanical resonators connected by mechanically coupled elements, as shown in Figure 8.1b. The summarized design parameters of devices 1 and 2 are presented in Table 8.1.

A finite element model (FEM) simulation is performed to demonstrate the vibration modes as shown in Figure 8.1c and d. Device 1 vibrates at flexural mode as given by Figure 8.1c, and a synchronization of mechanical elements affects device 2. Due to the limited PC memory, we just perform the FEM simulation for 44 resonant body beams and 43 mechanical coupled beams as shown in Figure 8.1d. The vibration synchronizes because of the mechanical interaction via the vibration of the coupling element connecting those resonators.

Each resonator is represented by a mass–spring–damper system, while the coupling beam corresponds to a network of mechanical springs. Its electrical equivalent circuit can be drawn as shown in Figure 8.2. The motional resistance R_m represents

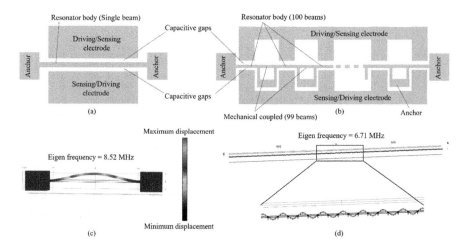

FIGURE 8.1   Device structures and FEM simulations. (a) Single capacitive silicon nanomechanical resonator. (b) Mechanical coupled capacitive silicon nanomechanical resonator.
(c) FEM simulation of the single capacitive resonator. (d) FEM simulation of the mechanical
coupled capacitive resonator.

TABLE 8.1
Summarized Parameters of Single and Mechanically Coupled
Capacitive Nanomechanical Silicon Resonators

Parameters	Single (Device 1)	Coupled (Device 2)
Width of resonant body (W)	500 nm	500 nm
Length of resonant body (L)	21.3 μm	21.3 μm
SOI wafer:	5 μm	5 μm
Device layer (t) (resonator height)	0.4 μm	0.4 μm
Buried SiO$_2$ layer	546 μm	546 μm
Silicon handling layer		
Capacitive gap (g)	300 nm	300 nm
Number of resonators	1	100
Number of couplings	0	99
V_{DC}	10 V	15 V
V_{AC}	0 dBm	0 dBm
Measured frequency (f_0)	9.9 MHz	7.2 MHz
Amplitude peak (A)	0.5 dB	20 dB
Quality factor (Q)	Nonlinear response	1000
Motional resistance (R_m)		1.2 kΩ

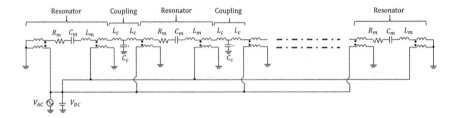

FIGURE 8.2 Electrical equivalent circuit of the mechanical coupled capacitive nanomechanical resonator.

mechanical losses of vibration. The motional inductance L_m indicates mechanical inertia. The motional capacitance C_m corresponds to mechanical compliance. The motional parameters of the single capacitive resonator without coupling can be calculated with the following equations [16–18]:

$$R_m = \frac{\sqrt{k_{eff} m_{eff}}}{Qn^2} = \frac{\sqrt{6.08 E \rho} g^4}{Q V_{DC}^2 \varepsilon_0^2} \frac{W^2}{L^3 t} \tag{8.1}$$

$$C_m = \frac{n^2}{k_{eff}} = \frac{V_{DC}^2 \varepsilon_0^2}{16 E g^4} \cdot \frac{L^5 t}{W^3} \tag{8.2}$$

$$L_m = \frac{m_{eff}}{n^2} = \frac{0.38 \rho g^4}{V_{DC}^2 \varepsilon_0^2} \cdot \frac{W}{L t} \tag{8.3}$$

where W is the width of the resonant body, L is the length of the resonant body, t is the thickness of the device layer, g is the capacitive gap width, ε_0 is the electric constant ($\varepsilon_0 = 8.854 \times 10^{-12}$ Fm^{-1}), and E and ρ are the Young's modulus and density of structure material, respectively. n presents the electromechanical conversion factor between the mechanical and electrical domain. k_{eff} and m_{eff} are the effective spring constant and mass, respectively.

$$n = V_{DC} \frac{\varepsilon_0 L t}{g^2} \tag{8.4}$$

$$k_{eff} = 16 E \frac{W^3 t}{L^3} \tag{8.5}$$

$$m_{eff} = 0.38 \rho L W t \tag{8.6}$$

where ρ is the density of the silicon material.

The motional resistance can be manipulated by increasing the polarization voltage V_{DC}, increasing the Q factor, decreasing the capacitive gap width, or increasing the overlap area of capacitance. Resonators can vibrate easily if the motional resistance is small and the motional capacitance is large. A simple method to achieve both requirements is increasing the electrode to resonator overlap area as shown in

Equations 8.1 and 8.2. Some research studies have investigated this issue, such as the capacitive coupling of bulk acoustic wave silicon resonators [19] or mechanically corner-coupled square resonator array [20]. The significant advantage of an array of coupled resonators is that it increases the overlap area provided over the individual resonator.

In turn, mechanical coupling of an nth number of resonators can be achieved by design; here each resonator is designed with the same parameters and fabricated at the same time with the same process. Each resonator is expected to have the same resonant frequency. Hence, lowering the effective motional resistance by the number of resonators can be achieved. The total motional resistance can be given by [20]

$$R_{\text{total}} = \frac{R_{\text{m}}}{n} \tag{8.7}$$

The resonant frequency f_0 of the clamped–clamped beam is calculated using the following equation:

$$f_0 = \frac{1}{2\pi} \sqrt{\frac{k_{\text{eff}}}{m_{\text{eff}}}} = 1.03 \sqrt{\frac{E}{\rho}} \frac{W}{L^2} \tag{8.8}$$

A resonant frequency by FEM simulation of device 1 is 8.52 MHz, while that of device 2 is 6.71 MHz. A possible reason for the difference is due to mode splitting. However, the resonators will still be vibrating in unison at the same frequency as they are mechanically coupled.

The coupling resonator can be considered as in Figure 8.3, where m is the effective mass of the resonator, and k_1 and k_2 are the spring constant of the resonant body and coupling, respectively.

The equation of motion x can be written as

$$m\ddot{x}_1 = -k_1 x_1 + k_2 (x_2 - x_1) = -(k_1 + k_2) x_1 + k_2 x_2$$

$$m\ddot{x}_2 = -k_1 x_2 - k_2 (x_2 - x_1) + k_2 (x_3 - x_2) = k_2 x_1 - (k_1 + 2k_2) x_2 + k_2 x_3$$

$$m\ddot{x}_3 = -k_1 x_3 - k_2 (x_3 - x_2) + k_2 (x_4 - x_3) = k_2 x_2 - (k_1 + 2k_2) x_3 + k_2 x_4$$

$$\cdots$$

$$m\ddot{x}_n = k_2 x_{n-1} - (k_1 + k_2) x_n.$$

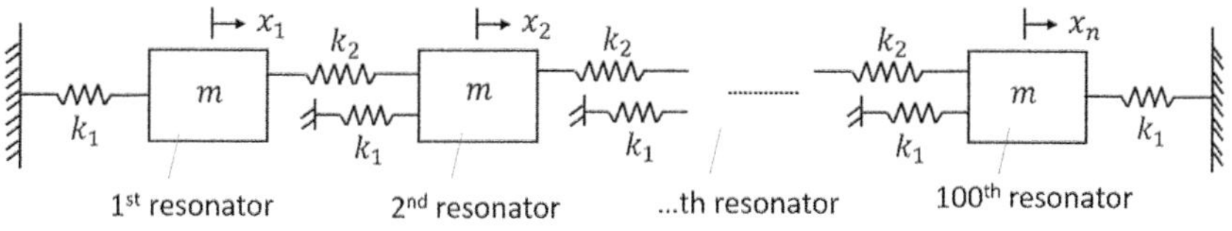

FIGURE 8.3 Coupling resonator.

The preceding equations can be considered as a z matrix:

$$m\begin{pmatrix} \ddot{x}_1 \\ \ddot{x}_2 \\ \ddot{x}_3 \\ \vdots \\ \ddot{x}_{n-1} \\ \ddot{x}_n \end{pmatrix}\begin{pmatrix} k_1 & -k_2 & 0 & & 0 & 0 \\ -k_2 & k_1+2k_2 & -k_2 & & 0 & 0 \\ 0 & -k_2 & k_1+2k_2 & & 0 & 0 \\ \vdots & \vdots & \vdots & \vdots & \cdots & \vdots & \vdots \\ 0 & 0 & 0 & & k_1+2k_2 & -k_2 \\ 0 & 0 & 0 & & -k_2 & k_1+k_2 \end{pmatrix}\begin{pmatrix} x_1 \\ x_2 \\ x_3 \\ \vdots \\ x_{n-1} \\ x_n \end{pmatrix}$$

which can be written as

$$m\ddot{X} = -KX \, (*).$$

If the number of resonators is small, the resonant frequency of each beam can be easily calculated by the diagonalization of the matrix. However, when the number of resonators is large (such as 100), calculation becomes difficult. In order to overcome the problem, we assume the fixing points at both sides are eliminated and connected to each other. It means the resonators are arranged in a ring. So the matrix can be easily diagonalized:

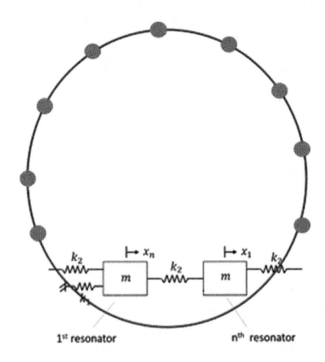

$$m\ddot{x}_1 = -k_2\left(x_1 - x_n\right) + k_2\left(x_2 - x_1\right) = -2k_2 x_1 + k_2 x_2 + k_2 x_n$$

$$m\ddot{x}_n = k_2\left(x_1 - x_n\right) - k_2\left(x_n - x_{n-1}\right) - k_1 x_n = k_2 x_1 + k_2 x_{n-1} - \left(k_2 + 2k_2\right)x_n$$

$$K = \begin{pmatrix} k_1 + 2k_2 & -k_2 & 0 & & 0 & -k_2 \\ -\ k_2 & k_1 + 2k_2 & -k_2 & & 0 & 0 \\ 0 & -k_2 & k_1 + 2k_2 & & 0 & 0 \\ \vdots & \vdots & \vdots & \cdots & \vdots & \vdots \\ 0 & 0 & 0 & & k_1 + 2k_2 & -k_2 \\ -k_2 & 0 & 0 & & -k_2 & k_1 + 2k_2 \end{pmatrix}$$

$$H = \begin{pmatrix} 0 & 1 & 0 & 0 & & 0 & 0 \\ 0 & 0 & 1 & 0 & & 0 & 0 \\ 0 & 0 & 0 & 1 & & 0 & 0 \\ \vdots & \vdots & \vdots & \vdots & \cdots & \vdots & \vdots \\ 0 & 0 & 0 & 0 & & 0 & 1 \\ 1 & 0 & 0 & 0 & & 0 & 0 \end{pmatrix}$$

Eigenvalues of the matrix are

$$\lambda_j = e^{i\frac{2\pi j}{n}}$$

with $j = 0, 1, 2, \ldots, n-1$.

Matrix diagonalization H is given:

$$U = \frac{1}{\sqrt{n}} \begin{pmatrix} 1 & 1 & 1 & & 1 \\ 1 & \lambda_1 & \lambda_2 & & \lambda_{n-1} \\ 1 & \left(\lambda_1\right)^2 & \left(\lambda_2\right)^2 & & \left(\lambda_{n-1}\right)^2 \\ \vdots & \vdots & \vdots & \cdots & \vdots \\ 1 & \left(\lambda_1\right)^{n-2} & \left(\lambda_2\right)^{n-2} & & \left(\lambda_{n-1}\right)^{n-2} \\ 1 & \left(\lambda_1\right)^{n-1} & \left(\lambda_2\right)^{n-1} & & \left(\lambda_{n-1}\right)^{n-1} \end{pmatrix}$$

$$U^{-1} = \frac{1}{\sqrt{n}} \begin{pmatrix} 1 & 1 & 1 & & 1 \\ 1 & (\lambda_1)^{n-1} & (\lambda_1)^{n-2} & & \lambda_1 \\ 1 & (\lambda_2)^{n-1} & (\lambda_2)^{n-2} & & \lambda_2 \\ \vdots & \vdots & \vdots & \cdots & \vdots \\ 1 & (\lambda_{n-2})^{n-1} & (\lambda_{n-2})^{n-2} & & \lambda_{n-2} \\ 1 & (\lambda_{n-1})^{n-1} & (\lambda_{n-1})^{n-2} & & \lambda_{n-1} \end{pmatrix}$$

So, we have

$$U^{-1}HU = \begin{pmatrix} 1 & 0 & 0 & 0 \\ 0 & \lambda_1 & 0 & 0 \\ 0 & 0 & \lambda_2 & 0 \\ \vdots & \vdots & \vdots & \cdots & \vdots \\ 0 & 0 & 0 & 0 \\ 0 & 0 & 0 & \lambda_{n-1} \end{pmatrix},$$

$$K = (k_1 + 2k_2)E - k_2H - k_2H^{n-1}$$

$K_d =$

$$\begin{pmatrix} k_1 & 0 & 0 & 0 \\ 0 & k_1 + 2k_2 - k_2\lambda_1 - k_2(\lambda_1)^{-1} & 0 & 0 \\ 0 & 0 & k_1 + 2k_2 - k_2\lambda_2 - k_2(\lambda_2)^{-1} & 0 \\ \vdots & \vdots & \vdots & \cdots & \vdots \\ 0 & 0 & 0 & 0 \\ 0 & 0 & 0 & k_1 + 2k_2 - k_2\lambda_{n-1} - k_2(\lambda_{n-1})^{-1} \end{pmatrix}$$

K_d is the diagonal matrix.

From earlier we have $m\ddot{X} = -KX$, which can be written as

$$m\frac{d^2}{dt^2}X = -KX$$

$$m\frac{d^2}{dt^2}X = -(UK_dU^{-1})X,$$

$$m\frac{d^2}{dt^2}(U^{-1}X) = -K_d(U^{-1}X)$$

$$m\frac{d^2}{dt^2}\left(U^{-1}X\right)=$$

$$-\begin{pmatrix} k_1 & 0 & & 0 & & 0 \\ 0 & k_1+2k_2-k_2\lambda_1-k_2\left(\lambda_1\right)^{-1} & & 0 & & 0 \\ 0 & 0 & & k_1+2k_2-k_2\lambda_2-k_2\left(\lambda_2\right)^{-1} & & 0 \\ \vdots & \vdots & & \vdots & \cdots & \vdots \\ 0 & 0 & & 0 & & 0 \\ 0 & 0 & & 0 & & k_1+2k_2-k_2\lambda_{n-1}-k_2\left(\lambda_{n-1}\right)^{-1} \end{pmatrix} U^{-1}X$$

We also have

$$\frac{d^2x}{dt^2}=-w^2x$$

where $x=A\cos\left(wt+\alpha\right)$

So that

$$U^{-1}X=\begin{bmatrix} A_1\cos\left(w_1t+\alpha_1\right) \\ \cdots \\ A_n\cos\left(w_nt+\alpha_n\right) \end{bmatrix}$$

$$\frac{d}{dt}\left(U^{-1}X\right)=\begin{bmatrix} -A_1w_1\sin\left(w_1t+\alpha_1\right) \\ \cdots \\ -A_nw_n\sin\left(w_nt+\alpha_n\right) \end{bmatrix}$$

$$\frac{d^2}{dt^2}\left(U^{-1}X\right)=\begin{bmatrix} -A_1w_1^2\cos\left(w_1t+\alpha_1\right) \\ \cdots \\ -A_nw_n^2\cos\left(w_nt+\alpha_n\right) \end{bmatrix}$$

So,

$$m\begin{bmatrix} -A_1w_1^2\cos\left(w_1t+\alpha_1\right) \\ \cdots \\ -A_nw_n^2\cos\left(w_nt+\alpha_n\right) \end{bmatrix}=K_d\begin{bmatrix} A_1\cos\left(w_1t+\alpha_1\right) \\ \cdots \\ A_n\cos\left(w_nt+\alpha_n\right) \end{bmatrix}$$

$$m\begin{bmatrix} w_1^2 \\ \cdots \\ w_n^2 \end{bmatrix}=K_d\begin{bmatrix} k_1 \\ \cdots \\ k_1+2k_2-k_2\lambda_{n-1}-k_2\left(\lambda_{n-1}\right)^{-1} \end{bmatrix}.$$

It means that

$$w_n^2 = \frac{1}{m}\left(k_1 + 2k_2 - k_2\lambda_{n-1} - k_2\left(\lambda_{n-1}\right)^{-1}\right)$$

$$w_n^2 = \frac{1}{m}(k_1 + 2k_2 - k_2\left(\lambda_{n-1} + \left(\lambda_{n-1}\right)^{-1}\right))$$

$$w_n^2 = \frac{1}{m}\left(k_1 + 2k_2 - 2k_2\cos\frac{2\pi j}{n}\right)$$

$$w_n^2 = \frac{1}{m}\left(k_1 + 2k_2\left(1 - \cos\frac{2\pi j}{n}\right)\right)$$

The final equation can be written as

$$w_n = \sqrt{\frac{1}{m}\left(k_1 + 2k_2\left(1 - \cos\frac{2\pi j}{n}\right)\right)} \text{ with } j = 0, 1, 2, \dots n-1. \qquad (8.9)$$

where m is the effective mass, and k_1 and k_2 are the spring constant of resonant body and coupling, respectively.

8.3 EXPERIMENTS

8.3.1 DEVICE FABRICATION

Resonator structures are fabricated by following the fabrication process shown in Figure 8.4. Details of the fabrication process are as follows.

An SOI wafer that consists of a 5 μm-thick top silicon layer, a 0.4 μm-thick oxide layer, and a 546 μm-thick silicon handling layer is employed (Figure 8.4a).

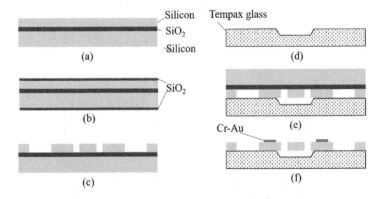

FIGURE 8.4 Fabrication process. (a) SOI wafer. (b) Thermal oxidation. (c) DRIE. (d) Wet etching. (e) Anodic bonding. (f) Handling and SiO_2 layer removal and Cr–Au electrodes.

A 300 nm-thick SiO_2 is grown on the SOI wafer by wet oxidation for the etching mask of the silicon device layer (Figure 8.4b). On this SiO_2 layer, an electron beam (EB) resist (ZEP 520A) pattern is formed by electron beam lithography. Then, the SiO_2 layer is etched by reactive ion etching (RIE) using a gas mixture of CHF_3 and Ar with a power of 120 W and a chamber pressure of 5 Pa. The capacitive gap and resonator structures are formed by using deep reactive ion etching (DRIE) by Bosch process using SF_6 and C_4F_8 for short etching and passivation cycles (etching time 2.5 sec, passivation time 2.5 sec) for creating low scallops (Figure 8.4c). The vertical sidewalls and small scallops of the capacitive gap are obtained as presented in Figure 8.5a. The resonant body width and capacitive gap size are approximately 500 nm and 300 nm, respectively, as shown in Figure 8.5a.

The resonator structures on the SOI wafer will be transferred to a glass substrate by anodic bonding in order to reduce the parasitic capacitances from the handling silicon layer [21]. A 300 µm-thick Tempax glass has been prepared for anodic bonding with the wafer. Before bonding, the top side of the Tempax glass is partly etched in diluted hydrofluoric acid (HF) using a metal mask (Cr–Au) (Figure 8.4d). The SOI wafer and the Tempax glass are aligned and bonded together at 400°C with

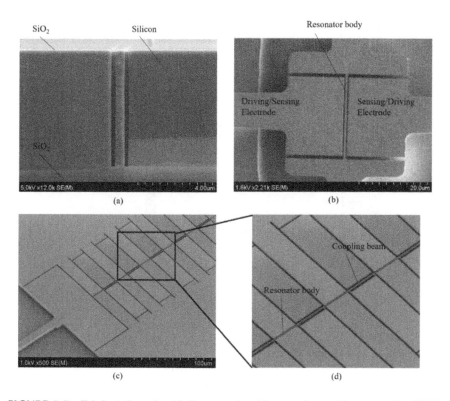

FIGURE 8.5 Fabricated results. (a) Cross-sectional image of capacitive gaps after DRIE. (b) Single capacitive silicon nanomechanical resonator. (c) Mechanically coupled capacitive silicon nanomechanical resonator. (d) Close-up image of mechanical coupled capacitive silicon resonator.

an applied voltage of 800 V for 15 min (Figure 8.4e). The handling silicon layer of the SOI wafer is removed by plasma etching using SF_6 gas. The buried SiO_2 layer is etched out by a buffered hydrofluoric acid (BHF) solution. Finally, the Cr–Au films with thicknesses of 20 and 300 nm, respectively, are partly deposited by a sputtering process via a stencil mask for electrode pads (Figure 8.4f). The fabricated devices 1 and 2 are shown in Figure 8.5b and c. Figure 8.5d shows a close-up image of the mechanically coupled capacitive silicon resonator.

8.3.2 MEASUREMENT RESULTS

The resonant characteristics of the fabricated devices 1 and 2 are evaluated, and the specifications are summarized in Table 8.1. Transmission S_{21} of device 1 is shown in Figure 8.6a for a single resonator with a length of 21.3 μm, width of 500 nm,

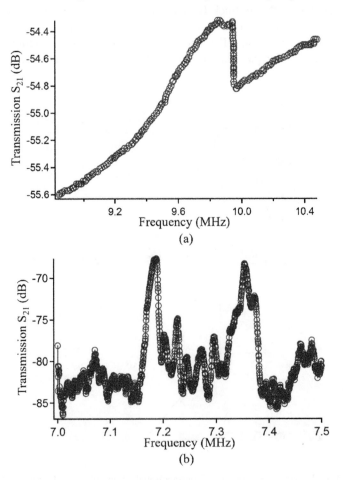

FIGURE 8.6 Frequency response of the fabricated devices. (a) Single capacitive silicon nanomechanical resonator. (b) Mechanically coupled capacitive silicon nanomechanical resonator.

thickness of 5 µm, and capacitive gap size of around 300 nm. A resonant peak, which is observed under measurement conditions $V_{DC}=10$ V and $V_{AC}=0$ dBm, is found at 9.9 MHz. The jump-and-drop phenomenon also arose in this device as shown in Figure 8.6a. The amplitude of frequency response increases as the frequency is swept upward, and then suddenly jumps to a lower value. This hard spring effect of beam resonators has also been observed [9, 22]. The displacement goes beyond the range of small perturbations; tension is applied to the resonating structure and this makes it stiffer. This situation is identical to the hardening spring.

The resonant peaks of device 2 are found at around 7.2 MHz. The resonant peaks are clearly observed with the resonant amplitude of 20 dB and a high Q factor of 1000 is achieved for 100 nanomechanical resonators . The observed resonant frequency of device 2 is lower than that of device 1. It is in good agreement with the FEM simulation as mentioned before. In order to evaluate the motional resistance of the coupling capacitive resonator, we assume the quality factor of coupling and single resonators are the same. Based on the L, W, t, V_{DC}, and assumed Q value and Equation 8.1, the motional resistance of the single capacitive resonator without coupling is about 120 kΩ. In turn, the motional resistance of 100 mechanical coupling resonators exhibits about 1.2 kΩ as calculated by Equation 8.7. Some resonant peaks have been observed (Figure 8.6b), which shows that most nanomechanical resonators are mechanically coupled and synchronized.

8.4 SUMMARY

We designed, fabricated, and evaluated mechanically coupled capacitive nanomechanical silicon resonators to detect small motional capacitance for attaining a low motional resistance for emerging sensing, image, and data processing technologies. Resonant peaks can be observed, which shows that most nanomechanical resonators are mechanically coupled and synchronized.

REFERENCES

1. van Beek, J.T.M.V., Puers, R., A review of MEMS oscillators for frequency reference and timing applications, *Journal of Micromechanics and Microengineering*, **22**, 013001, 2012.
2. Toan, N.V., Miyashita, H., Toda, M., Kawai, Y., Ono, T., Fabrication of an hermetically packaged silicon resonator on LTCC substrate, *Microsystem Technologies*, **19**, 1165–1175, 2013.
3. Gieseler, J., Novotny, L., Quidant, R., Thermal nonlinearities in a nanomechanical oscillator, *Nature Physics*, **9**, 806–810, 2013.
4. Chaste, J., Eichler, A., Moser, J., Ceballos, G., Rurali, R., Bachtold, A., A nanomechanical mass sensor with yoctogram resolution, *Nature Nanotechnology*, **7**, 301–304, 2012.
5. Seo, Y.J., Toda, M., Ono, T., Si nanowire probe with Nd-Fe-B magnet for attonewton scale force detection, *Journal of Micromechanics and Microengineering*, **25**, 045015, 2015.
6. Inomata, N., Toda, M., Sato, M., Ishijima, A., Ono, T., Pico calorimeter for detection of heat produced in an individual brown fat cell, *Applied Physics Letters*, **100**, 154104, 2012.

7. Rips, S., Hartmann, M.J., Quantum information processing with nanomechanical qubits, *Physical Review Letters*, **111**, 049905, 2013.
8. Bannon, F.D., Clark, J.R., Nguyen, C.T.C., High Q factor HF micromechanical filters, *Journal of Solid-State Circuits*, **35**, 512–526, 2000.
9. Mestron, R.M.C., Fey, R.H.B., Phan, K.L., Nijmeijer, H., Experimental validation of hardening and softening resonances in a clamped-clamped beam MEMS resonator, *Proceedings of the Erosensor XXIII Conference*, 812–815, 2009.
10. Toan, N.V., Kubota, T., Sekhar, H., Samukawa, S., Ono, T., Mechanical quality factor enhancement in a silicon micromechanical resonator by low-damage process using neutral beam etching technology, *Journal of Micromechanics and Microengineering*, **24**, 085005, 2014.
11. Toan, N.V., Toda, M., Kawai, Y., Ono, T., A long bar type silicon resonator with a high quality factor, *IEEJ Transactions on Sensors and Micromachines*, **134**, 26–31, 2014.
12. Lin, Y.W., Li, S.S., Xie, Y., Ren, Z., Nguyen, C.T.C., Vibrating micromechanical resonators with solid dielectric capacitive transducers gaps, *Proceedings of IEEE Frequency Control Symposium and Exposition*, 128–134, 2005.
13. Toan, N.V., Toda, M., Kawai, Y., Ono, T., A capacitive silicon resonator with movable electrode structure for gap width reduction, *Journal of Micromechanics and Microengineering*, **24**, 025006, 2014.
14. Pourkamali, S., Ho, G.K., Ayazi, F., Low impedance VHF and UHF capacitive silicon bulk acoustic wave resonators – Part I: Concept and fabrication, *IEEE Transactions on Electron Devices*, **54**, 2017–2023, 2007.
15. Sage, E., Martin, O., Dupre, C., Ernst, T., Billiot, G., Duraffourg, L., Colinet, E., Hentz, S., Frequency addressed NEMS arrays for mass and gas sensing applications, *Proceedings of Transducers Conference*, 665–668, 2013.
16. Pourkamali, S., Hashimura, A., Abdolvand, R., Ho, G.K., Erbil, A., Ayazi, F., High Q single crystal silicon HARPSS capacitive beam resonators with self-aligned sub-100 nm transduction gap, *Journal of Microelectromechanical Systems*, **12**, 487–496, 2003.
17. Jin, Y., Tang, Y., Yu, X., Study on clamped-clamped beams in-plane capacitive resonators, *Proceedings of SPIE*, **7159**, 71590P-1–71590P-8, 2009.
18. Durand, C., Casset, F., Ancey, P., Judon, F., Talbot, A., Quenouillere, R., Renaund, D., Borel, S., Florin, B., Buchaillot, L., Silicon on nothing MEMS electromechanical resonator, *Design Test Integration Packaging*, 2007.
19. Qishu, Q., Pourkamali, S., Ayazi, F., Capacitively coupled VHF silicon bulk acoustic wave filters, *IEEE Ultrasonics Symposium*, 1649–1652, 2007.
20. Demirci, M.U., Nguyen, C.T.-C., Mechanically corner-coupled square microresonator array for reduced series motional resistance, *Journal of Microelectromechanical Systems*, **15**, 1419–1436, 2006.
21. Okada, M., Nagasaki, H., Tamano, A., Niki, K., Tanigawa, H., Suzuki, K., Silicon beam resonator utilizing the third-order bending mode, *Japanese Journal of Applied Physics*, **48**, 06FK03, 2009.
22. Husain, A., Hone, J., Postma, H.W.Ch., Huang, X.M.H., Drake, T., Barbic, M., Scherer, A., Roukes, M.L., Nanowire-based very high frequency electromechanical resonator, *Applied Physics Letters*, **83**, 1240–1242, 2003.

9 Capacitive Silicon Nanomechanical Resonators with Selective Vibration of High-Order Mode

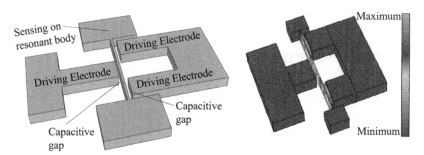

Sensing on resonant body

Driving Electrode

Driving Electrode

Driving Electrode

Capacitive gap

Capacitive gap

Maximum

Minimum

3rd mode vibration structure   FEM simulation of 3rd mode vibration structure

9.1 INTRODUCTION

Many Internet of things (IoT) devices can be connected to the Internet via wireless network [1]. Increasing the amount of the transmitted and received information and accurate data transmissions are necessary. Satisfying these issues, micro clock generators for transmitters and receivers with a smaller size and higher performance are required.

Quartz crystal resonators are usually employed for the aforementioned applications, which exhibit a high quality factor [2, 3], good power handling [4], and an excellent temperature stability [5, 6]. However, their vibration is on a small scale owing to direct physical contact between electrodes and the resonant body. Their working frequencies depend on the thickness of the piezoelectric film, which has a low range frequency. Also, their fabrication process is not compatible with complementary metal–oxide semiconductor (CMOS) fabrication. Capacitive silicon resonators, on the other hand, are expected to overcome the noted problems, as presented in many works [7–13]. An ultra-high Q factor can be achieved by the capacitive silicon resonators [7–9]. In addition, they are capable of integration with a CMOS chip [10] as well as excellent long-term stability [7, 8]. Their resonant frequency (fundamental mode) depends on the geometric dimensions of the resonators [7, 13]. For instance, resonant frequencies of bar-type [13], square-type [12], and disk-type [9] capacitive silicon resonators are decided by the width of the resonant body. A fixed–fixed beam capacitive silicon resonator presented by van Beek and Puers [7] contains a resonant beam body, as well as driving and sensing electrodes. Its resonant frequency is determined by its length and width. In attaining high-frequency capacitive silicon resonators, downscaling (reducing the length and width of the resonant body) is the common solution, which induces problems such large motional resistance and high insertion loss. The motional resistance of the resonators is always desired to be as low as possible for an impedance matching with the CMOS chip. Hence, the downscaling method makes the capacitive silicon resonators difficult for practical applications (i.e., integration with electrical circuits). Pursuant to an increase of the operating frequency without downscaling the resonant structures, this work focuses on capacitive silicon nanomechanical resonators able to vibrate at a higher mode selectively based on placing the driving electrodes along the resonant body. The 1st, 2nd, 3rd, and 4th mode fixed–fixed beam capacitive silicon resonators are produced and examined.

9.2 DEVICE DESCRIPTION

Figure 9.1a and b present a perspective-view schematic of the fixed–fixed beam capacitive silicon resonators with 1st and 3rd mode vibration, respectively. The basic components of resonator structures are the resonant body, capacitive gaps, the driving electrode, and the sensing electrode. The resonant body is suspended by two

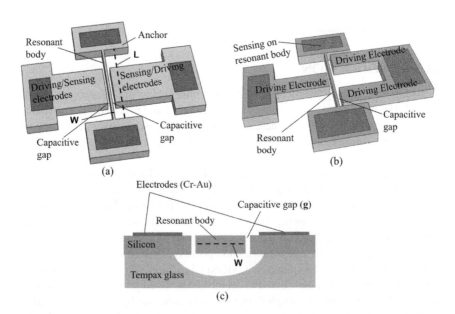

FIGURE 9.1 Fixed–fixed beam capacitive silicon resonators. (a) 1st mode vibration structure. (b) 3rd mode vibration structure. (c) Cross-sectional structure.

anchors at both ends of the resonant body on patterned glass substrate. The cross-sectional view of the resonator structure is shown in Figure 9.1c. For the 1st mode vibration structure, the driving electrode is placed along the side of the resonant body and the sensing electrode is on the opposite side, as shown in Figure 9.1a. They are separated from the resonant body by narrow capacitive gaps. However in 3rd mode vibration, the driving electrodes are designed and placed on both sides of the resonant body, and the sensing electrode for motional detection is on the resonant body electrode (Figure 9.1b). High-order mode vibration structures are defined by the number of driving electrodes along the resonant body. The number of driving electrodes for the 2nd, 3rd, and 4th mode vibration are 2, 3, and 4, respectively.

To operate the resonators, an alternating current (AC) input signal V_{AC} together with a direct current (DC) bias voltage V_{DC} are applied to a driving electrode that results in an electrostatic force acting on the resonant body vibration. This motion results in changes of the motional capacitance of the resonators owing to the changes in the size of the capacitive gaps. Based on monitoring in a time-varying electrostatic force, the resonant frequency of the resonators can be observed.

The resonant body is actuated by a delta deviance electrostatic force, which is generated by the combined effects of DC voltage and AC voltage, given as

$$\Delta F = \frac{\varepsilon_r \varepsilon_0 A_{el}}{g^2} V_{DC} V_{AC}, \tag{9.1}$$

where A_{el} is the area of the electrode plate, ε_r is the dielectric constant of the material between the plates (for an air environment, $\varepsilon_r \approx 1$), ε_0 is the electric constant ($\varepsilon_0 \approx 8.854 \times 10^{-12} \, \mathrm{Fm^{-1}}$), and g is the distance between two plates called the capacitive gap.

The resonant frequencies f_n are determined by the formula of the effective spring constant k_{eff} and the effective mass m_{eff}, as follows:

$$f_n = \frac{1}{2\pi} \sqrt{\frac{k_{eff}}{m_{eff}}}. \tag{9.2}$$

The effective spring constant and effective mass of the resonators are given by

$$k_{eff} = \lambda_n^4 n \frac{EI_z}{L^3}, \tag{9.3}$$

$$m_{eff} = n m_0, \tag{9.4}$$

where λ_n is the frequency coefficient for each resonance mode, E is Young's modulus of the resonator material, I_z is the area moment of inertia, l is the length of the resonant body, and m_0 is the mass of the resonators.

Equations 9.2, 9.3, and 9.4 can be combined as [14]

$$f_n = k_n \frac{W}{L} \sqrt{\frac{E}{\rho}}, \tag{9.5}$$

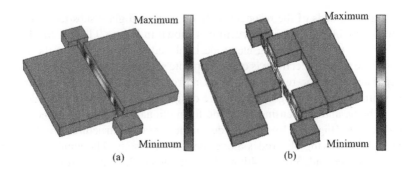

FIGURE 9.2 FEM simulation. (a) 1st mode vibration. (b) 3rd mode vibration.

where k_n is the corresponding constant value for each resonance mode and ρ is the density of the structure material. The k_n values for the 1st, 2nd, 3rd, and 4th resonance modes are $k_1 = 1.027$, $k_2 = 2.833$, $k_3 = 5.54$, and $k_4 = 9.182$, respectively.

The motional resistance R_m represents the mechanical loss of the vibration and can be extracted from the insertion loss as follow [15]:

$$R_m = 50 \left(10^{\frac{IL_{dB}}{20}} - 1 \right), \tag{9.6}$$

where IL_{dB} is the insertion loss of the transmission and its unit is in decibels (dB).

Fixed–fixed beam resonators presented in this work are designed for lateral vibration as a flexural mode. A finite element method (FEM) model is built by COMSOL for a prediction of the vibration shape and the resonant frequency. Figure 9.2a and b show the vibration shapes of the 1st and 3rd modes, respectively. The colors correspond to total in-plane displacement where the white gray is the maximum displacement and dark gray is no displacement. Other vibration mode shapes can be found in Table 9.1.

In this work, all resonators are designed to have the same resonant frequency for performance comparison. Fixed–fixed beam resonators with 1st, 2nd, 3rd, and 4th mode vibrations are designed and fabricated. The resonator parameters, their theoretical calculation, and FEM simulation are shown in Table 9.1.

9.3 EXPERIMENTS

In this section, the fabrication process, measurement setups, and measurement results are presented. Devices are produced by electron beam lithography (EBL), photolithography, deep reactive ion etching (RIE), and anodic bonding. To save the time taken during the exposing process of EBL, only nano capacitive gaps are formed. An extra conventional photolithography is subsequently performed to create the resonator structures. The silicon resonator structures formed on a silicon on insulator (SOI) wafer are transferred to a glass substrate by anodic bonding in order to reduce the parasitic capacitances from the handling silicon layer. The detailed information is as follows.

TABLE 9.1

Summary of Parameters of the 1st, 2nd, 3rd, and 4th Mode Capacitive Resonators

	1st Mode	2nd Mode	3rd Mode	4th Mode
		Parameters		
Resonant length	21.3 μm	35.5 μm	49.3 μm	63.3 μm
Resonant width	0.5 μm	0.5 μm	0.5 μm	0.5 μm
Resonant thickness	7 μm	7 μm	7 μm	7 μm
Capacitive gap	0.3 μm	0.3 μm	0.3 μm	0.3 μm
Number of driving electrode	1	2	3	4
		Calculation		
Frequency	9.66 MHz	9.60 MHz	9.73 MHz	9.79 MHz
		FEM Simulation		
Frequency	9.71 MHz	9.68 MHz	9.73 MHz	9.78 MHz
Vibration mode (resonant body only)				
		Measurement Conditions		
V_{AC}	0 dBm	0 dBm	0 d Bm	0 dBm
V_{DC}	15 V	15 V	15 V	15 V
Pressure level	0.01 Pa	0.01 Pa	0.01 Pa	0.01 Pa
		Experimental Results		
Resonant frequency	10.15 MHz	10.85 MHz	10.85 MHz	10.36 MHz
Quality factor	10078	8768	4255	844
Insertion loss	−75 dB	−71 dB	−65.6 dB	−51.5 dB
Motional resistance	281 kΩ	181 kΩ	95 kΩ	18.7 kΩ

9.3.1 EXPERIMENTAL METHODOLOGY

The starting substrate is the SOI wafer with a 7 μm-thick device layer with a low resistivity of 0.02 Ωcm, a 1 μm-thick oxide layer, and a 300 μm-thick silicon handling layer (Figure 9.3a). After conventional cleaning, including RCA1, RCA2, and piranha, an approximate 500 nm-thick SiO_2 layer is formed on the entire surface of the SOI wafer using wet thermal oxidation (Figure 9.3b). Then, a 400 nm-thick EB resist (ZEP 502A) is patterned on the above SiO_2 layer (on the device layer side). The reactive ion etching (RIE) method is employed to etch SiO_2 with the EB resist as a mask and using a gas mixture of CHF_3 and Ar with a power of 120 W and a chamber pressure of 5 Pa. Narrow gaps with smooth and vertical-etched shapes were achieved, as shown in Figure 9.4a. After removing the EB resist, the nanotrenches on the top silicon layer are then formed by DRIE with the Bosch process using SF_6 (etching cycles of 2.5 sec) and C_4F_8 (passivation cycles of 2.5 sec) gases. Figure 9.4b shows the formed nanotrenches using the aforementioned process. The resonant body and capacitive gaps are 500 nm and 300 nm, respectively. Following this, the resonator structures are created by photolithography after the DRIE of silicon (Figure 9.3c).

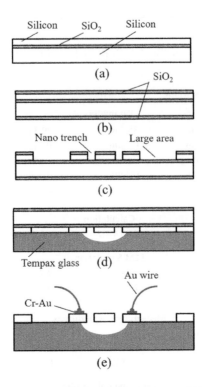

FIGURE 9.3 Fabrication process. (a) SOI wafer (7 μm/1 μm/300 μm). (b) Thermal oxidation. (c) Combination of EBL, photolithography, and DRIE process. (d) Anodic bonding. (e) Back side silicon etching, SiO₂ removal, and metal contact pads.

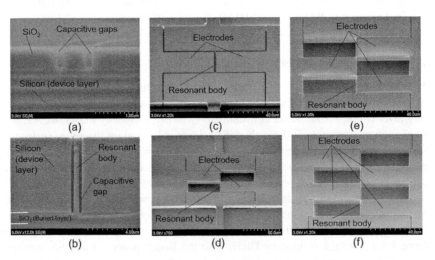

FIGURE 9.4 Fabricated results. (a) SiO₂ patterning with EB resist and using RIE technique. (b) Resonant body and narrow trenches formed by DRIE. (c) 1st mode vibration structure. (d) 2nd mode vibration structure. (e) 3rd mode vibration structure. (f) 4th mode vibration structure.

To reduce the parasitic capacitances from the handling silicon substrate, the 300 μm-thick Tempax glass substrate is employed for the transferring process. The resonator structures on the SOI wafer are aligned and bonded with the Tempax glass substrate by the anodic bonding technique (Figure 9.3d). The bonding process is performed at 400°C with 800 V power sources for 15 min. The back side silicon handling layer is removed by plasma etching of SF_6 gas. After the buried SiO_2 layer is etched out by buffered hydrofluoric acid (BHF) solution and devices are dried by a supercritical CO_2 process to avoid sticking issues, the electrode pads using Cr–Au are formed by a sputtering process via a shadow mask. Finally, Au wire bonding is conducted, as shown in Figure 9.3e.

Figure 9.4c–f shows the successfully fabricated devices with differently expected mode vibration shapes of the 1st, 2nd, 3rd, and 4th mode, respectively.

9.3.2 Measurement Setup

The frequency responses of the fabricated devices are evaluated by the electrical setup shown in Figure 9.5. The measurement setup contains a network analyzer (E5071B ENA Series, Agilent Technologies) with a frequency range from 300 kHz to 8.5 GHz, DC voltage source V_{DC}, electrical components including capacitors and resistors, and coaxial cables. The resonators are set in a vacuum chamber at a pressure chamber of 0.01 Pa.

To detect the motion of the 1st mode vibration structure, the setup shown in Figure 9.5a is employed. Output and input ports of the network analyzer are connected to the driving and sensing electrodes, respectively, through the resistors and capacitors, while the resonant body electrode is attached to the ground. The purpose of usage of the resistors and capacitors is to decouple the radio frequency (RF) signal and also to avoid damage to the network analyzer. With the high-order mode vibration, all driving electrodes on both sides of the resonant body electrode are connected to the output port of the network analyzer, while the resonant body is connected to the input port of the network analyzer (Figure 9.5b).

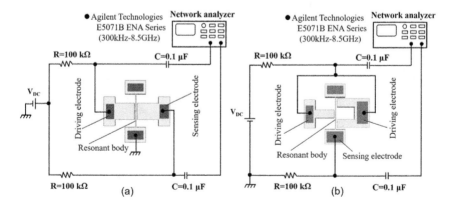

FIGURE 9.5 Measurement setups. (a) 1st mode vibration structure. (b) High-order mode vibration structure.

9.3.3 MEASUREMENT RESULTS

Specifications of the fabricated devices are summarized in Table 9.1. Frequency responses of the fabricated devices are shown in Figure 9.6. Evaluation conditions are the same for all fabricated devices under $V_{DC} = 15$ V and $V_{AC} = 0$ dBm and a vacuum chamber of 0.01 Pa. Similar resonant frequency values are observed for all fabricated devices although their resonant lengths are significantly different (Table 9.1). Thus, placing the driving electrodes along the resonant body, the high-order mode capacitive resonators can be demonstrated.

The resonant peak of the 1st mode vibration structure is found at 10.15 MHz with a quality factor Q of 10,000 as shown in Figure 9.6a. The measured resonant frequency is in good agreement with the FEM simulation result (Table 9.1). Figure 9.6b, c, and d shows the frequency responses of the 2nd, 3rd, and 4th mode vibration structures, respectively. No other vibration modes have been observed in these resonators.

The quality factor decreases as the vibration mode increases. The possible reason is due to the large supporting loss [16, 17] and thermoplastic damping [18–20] of the high-order mode compared to that of the lower modes, which result in a high energy dissipation. Nevertheless, the insertion loss of the fabricated devices is improved from –75 dB to –51.5 dB as the vibration modes are raised from the 1st to 4th mode capacitive devices. The motional resistance of the 1st mode vibration structure is 281 kΩ,

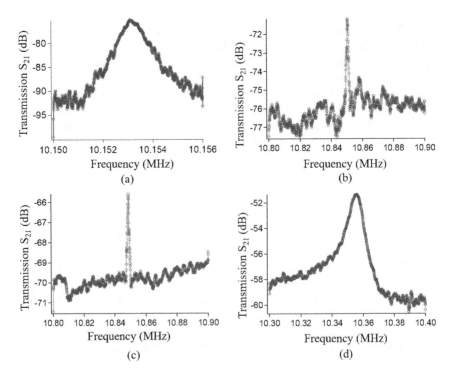

FIGURE 9.6 Frequency responses. (a) 1st mode vibration structure. (b) 2nd mode vibration structure. (c) 3rd mode vibration structure. (d) 4th mode vibration structure.

whereas those of the 2nd, 3rd, and 4th mode vibration structures are 181, 95, and 18.7 kΩ, respectively. Thus, the motional resistances of the 4th mode vibration structure are reduced by 83%, 90%, and 93% over the 3rd, 2nd, and 1st mode vibration structures, respectively.

The methods to reduce the motional resistance while maintaining the high Q factor comparable to that of the first mode vibration structure are suggested as follows: low supporting loss by U-shaped supports instead of straight supports [21], low damage etched surfaces by the choice of fabrication technologies [22], and low thermoplastic dissipation by optimizing the design [12].

9.4 SUMMARY

In this chapter, the high-order mode capacitive silicon resonators with the driving electrodes placed along the resonant body are produced and examined. It is demonstrated that the high-order mode resonators can achieve lower insertion loss and smaller motional resistance over those of the low-order mode resonators. Fixed–fixed beam capacitive silicon resonators as well as other types of capacitive silicon resonators, including bar type, disk type, and square type, could be employed in this proposal for the high-order vibration modes with low motional resistance.

REFERENCES

1. IoT devices and local networks. Available online: https://www.micrium.com/iot/devices/ (accessed on September 11, 2017).
2. Gallio, S., Goryachev, M., Abbe, P., Vacheret, X., Tobar, M.E., Bourquin, R., Quality factor measurement of various type of quartz crystal resonator operating near 4K, *IEEE Transactions on Ultrasonics, Ferroelectrics, and Frequency Control*, **63**, 975–980, 2016.
3. Suzuki, R., Sakamoto, K., Watanabe, Y., Method for improving quality factor in crystal oscillators with duplicated quartz resonator, *Proceedings of Symposium on Ultrasonic Electronics*, **34**, 259–260, 2013.
4. Murizaki, Y., Sakuma, S., Arai, F., Microfabrication of wide-measurement-range load sensor using quartz crystal resonator, *28th IEEE International Conference on Micro Electro Mechanical Systems*, Estoril, Portugal, 833–836, 2015.
5. Spassov, L., Gadjanova, V., Velcheva, R., Dulmet, B., Short and long term stability of resonant quartz temperature sensors, *IEEE Transactions on Ultrasonics, Ferroelectrics, and Frequency Control*, **55**, 1626–1631, 2008.
6. Murozaki, Y., Arai, F., Wide range load sensor using quartz crystal resonator for detection of biological signals, *IEEE Sensors Journal*, **15**, 1913–1919, 2015.
7. van Beek, J.T.M., Puers, R., A review of MEMS oscillators for frequency reference and timing applications, *Journal of Micromechanics and Microengineering*, **22**, 013001, 2012.
8. Nguyen, C.T.C., MEMS technology for timing and frequency control, *IEEE Transactions on Ultrasonics, Ferroelectrics, and Frequency Control*, **54**, 251–270, 2007.
9. Abdolvand, R., Bahreyni, B., Lee, J.E.Y., Nabki, F., Micromachined resonators: A review, *Micromachines*, **7**, 60, 2016.
10. Toan, N.V., Miyashita, H., Toda, M., Kawai, Y., Ono, T., Fabrication of an hermetically packaged silicon resonator on LTCC substrate, *Microsystem Technologies*, **19**, 1165–1175, 2013.

11. Toan, N.V., Toda, M., Kawai, Y., Ono, T., Capacitive silicon resonator with movable electrode structure for gap width reduction, *Journal of Micromechanics and Microengineering*, **24**, 025006, 2014.

12. Toan, N.V., Nha, N.V., Song, Y., Ono, T., Fabrication and evaluation of capacitive silicon resonators with piezoresistive heat engines, *Sensors and Actuators A: Physical*, **262**, 99–107, 2017.

13. Toan, N.V., Toda, M., Kawai, Y., Ono, T., A long bar type silicon resonator with a high quality factor, *IEEJ Transactions on Sensors and Micromachines*, **134**, 26–31, 2014.

14. Bahreyni, B., *Fabrication and Design of Resonant Microdevices*, William Andrew Inc., Norwich, NY, 2008.

15. Ho, G.K., Sundaresan, K., Pourkamali, S., Ayazi, F., Low motional impedance highly tunable I2 resonators for temperature compensated reference oscillators, *18th IEEE International Conference on Micro Electro Mechanical Systems*, Miami Beach, FL, 116–120, 2005.

16. Hao, Z., Erbil, A., Ayazi, F., An analytical model for support loss in micromachined beam resonators with in-plane flexural vibrations, *Sensors and Actuators A: Physical*, **109**, 156–164, 2003.

17. Yasumura, K.Y., Stowe, T.D., Chow, E.M., Pfafman, T., Kenny, T.W., Stipe, B.C., Rugar, D., Quality factor in micron- and submicron-thick cantilevers, *Journal of Microelectromechanical Systems*, **9**, 117–125, 2000.

18. Lifshitz, R., Roukes, M.L., Thermoelastic damping in micro and nanomechanical systems, *Physical Review B*, **61**, 5600–5609, 2000.

19. Duwel, A., Weinstein, M., Gorman, J., Borenstein, J., Ward, P., Quality factors of MEMS gyros and the role of thermoelastic damping, *Proceedings of the 15th IEEE International Conference on MEMS*, Las Vegas, NV, 214–219, 2000.

20. Pourkamali, S., Hashimura, A., Abdolvand, R., Ho, G.K., Erbil, A., Ayazi, F., High Q single crystal silicon HARPSS capacitive beam resonator with self-aligned sub-100 nm transduction gaps, *Journal of Microelectromechanical Systems*, **12**, 487–496, 2003.

21. Lee, J.E.Y., Yan, J., Seshia, A.A., Low loss HF band SOI wine glass bulk mode capacitive square-plate resonator, *Journal of Micromechanics and Microengineering*, **19**, 074003, 2009.

22. Toan, N.V., Kubota, T., Sekhar, H., Samukawa, S., Ono, T., Mechanical quality factor enhancement in a silicon micromechanical resonator by low-damage process using neutral beam etching technology, *Journal of Micromechanics and Microengineering*, **24**, 085005, 2014.

10 Capacitive Silicon Resonators with Movable Electrode Structures

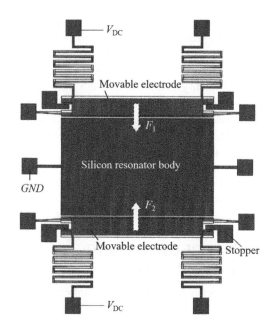

10.1 INTRODUCTION

There are many ways to reduce motional resistance, such as increasing the polarization voltage V_{DC}, increasing the overlap area of the capacitance, increasing the Q factor, or decreasing the capacitive gap width.

Increasing the polarization voltage V_{DC} over the gap is not effective because a high voltage is not currently available on a complementary metal–oxide semiconductor (CMOS) chip. It means that more challenges and difficulties face the integration process between silicon resonators and large-scale integration (LSI). When applying a high voltage to the resonant structure, and if the electrostatic force overcomes the stiffness constant of the resonator structure, the pull-in phenomenon occurs, which makes the resonant body and electrode contact. Increasing the electrode area, using resonator arrays [1–6], such as capacitive coupling of bulk acoustic wave silicon resonators or mechanically corner-coupled square resonator arrays, can reduce the motional resistance and increase the power handling. However, the frequency response and the Q factor of the silicon resonator array suffer from the mismatch of the resonant frequencies. To increase the quality factor, it is necessary to consider the loss factors including external loss and internal loss [7]. Nevertheless, the motional resistance only depends on the first order of the Q factor. The gap reduction is considered as one of the best methods since the motional resistance is proportional to the fourth order of the capacitive gap width.

Many researchers have been performed experiments to reduce the gap width, for instance, resonant structures with a solid filled electrode on the resonator gap [8, 9]. A high dielectric constant thin film (silicon nitride), which is used to replace the air gap of conventional resonant structures on the dielectric film, is sandwiched between the electrodes and the resonant body. The motional resistance of these structures had small value because of the high dielectric constant and very narrow solid gap (15 nm silicon nitride [8] and 20 nm silicon nitride [9]). However, the resonant structure suffers from material mismatch between the silicon and silicon nitride material, which decreases the mechanical quality factor due to high internal loss of the interface losses induced by the deposited silicon nitride film [7]. A fabrication method of resonators with small capacitive gap using a thin oxide film as a sacrificial layer has also been attempted [4, 10–13]. However, this approach is slightly complex and a high temperature process is required.

In this chapter, a simple fabrication process for a narrow gap using direct etching is proposed. In addition, in order to compensate for temperature drifts of oscillation frequency by an electrostatic tuning method, a capacitive resonator with movable electrodes is proposed.

10.2 FUNDAMENTALS OF ELECTROSTATIC PARALLEL PLATE ACTUATION

There are many kinds of electrostatic actuators, including the curved electrode [14], torsional ratcheting actuator [15–17], com-drive actuator [18, 19], and parallel plate actuator [20–22]. Each of these actuators has its unique benefits, applications, and limitations. In this section, the fundamentals of electrostatic parallel plate actuation will be briefly explained. The electrostatic parallel plate actuation, a parallel plate capacitor, has been widely used for radio-frequency microelectromechanical systems (RF MEMS), resonators, mirrors, and switches [20–22]. This capacitor consists of movable and fixed electrodes, and the movable electrode moves straight along the electric field paths. Generally, the electrostatic actuators operate in the basic principle of like charges repel and opposite charges attract. These devices utilize columbic attraction between two bodies to induce displacement or exert force.

The electrostatic parallel plate actuator consists of two parallel plate electrodes that are formed with fixed and movable electrodes as in Figure 10.1. The operation of this structure is as follows. The electrostatic force between the fixed and movable electrodes is zero when the voltage across the electrodes is not applied ($V=0$). When the voltage between two electrodes is increased, the resulting electrostatic force between two electrodes pulls the movable electrode toward the fixed electrode until the electrostatic force F_E equals the restoring force F_R. The restoring force F_R

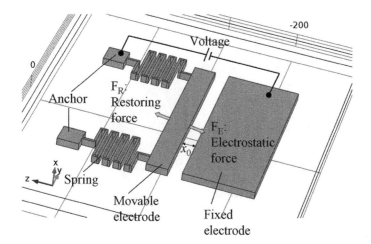

FIGURE 10.1 Schematic of two parallel-plate electrodes before a voltage is applied.

is created from elastic spring stiffness, which acts to bring the movable electrode back toward equilibrium as in Figure 10.1. The electrostatic force that depends on the voltage is acting in the opposite direction from the elastic restoring spring force. The restoring force is effectively reduced, so the movable electrode acts as though it has a reduced spring constant with increasing polarization voltage.

The electrostatic force is calculated from the stored energy E_S in the electrostatic field and it is shown as

$$E_S = \frac{1}{2}CV^2, \tag{10.1}$$

where C and V are the capacitance and the voltage, respectively, across the capacitor. The capacitance between the electrodes is given by

$$C = \frac{\varepsilon_0 \varepsilon_r A}{d}, \tag{10.2}$$

where ε_0 is the permittivity of dielectric (8.854×10^{-12} Fm^{-1}), ε_r is the relative permittivity, A is the area of the electrode, and d is the separation distance of two movable electrodes. Equation 10.2 is substituted into Equation 10.1, and the stored energy can be rewritten as

$$E_S = \frac{\varepsilon_0 \varepsilon_r A V^2}{2d}. \tag{10.3}$$

As shown in Figure 10.2, if the movable electrode is moved approximately x from initial distance x_0 between two electrodes, the separation distance is expressed as

$$d = x_0 - x, \tag{10.4}$$

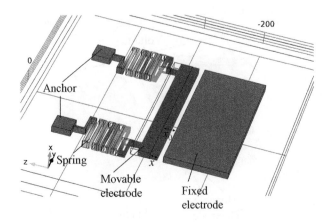

FIGURE 10.2 Schematic of two parallel-plate electrodes after a voltage is applied.

and the stored energy is given by

$$E_S = \frac{\varepsilon_0 \varepsilon_r A V^2}{2(x_0 - x)},$$
(10.5)

If the movable electrode is allowed to move an infinitesimally small distance Δx, the actuator will gain energy when the voltage across the electrodes is held constant. Thus, in this case, an electrical charge at any point in an electric field possesses the potential energy E_P, and it interacts with the positive and negative charges. This electrical potential energy is generated due to the position of the charge relative to other charges.

Therefore, the energy balance in an electric field is expressed as

$$F_E \Delta x = \frac{dE_S}{dx} \Delta x - \frac{dE_P}{dx} \Delta x.$$
(10.6)

Here, $F_E \Delta x$ equals the energy gained by the capacitance between electrodes minus the internal energy lost by the power supply. Equation 10.6 can briefly be rewritten as

$$F_E = \frac{dE_S}{dx} - \frac{dE_P}{dx},$$
(10.7)

where

$$\frac{dE_S}{dx} = \frac{V^2 \partial C(x)}{2 \partial x} = -\frac{\varepsilon_0 \varepsilon_r A V^2}{2(x_0 - x)^2},$$
(10.8)

$$q(x) = C(x)V,$$
(10.9)

$$\frac{dE_P}{dx} = \frac{\partial q(x)}{\partial x} V = \frac{V^2 \partial C(x)}{\partial x} = -\frac{\varepsilon_0 \varepsilon_r A V^2}{(x_0 - x)^2}.$$
(10.10)

Therefore, the electrostatic force is given by

$$F_E = -\frac{\varepsilon_0 \varepsilon_r A V^2}{2(x_0 - x)^2} - (-\frac{\varepsilon_0 \varepsilon_r A V^2}{(x_0 - x)^2}) = \frac{\varepsilon_0 \varepsilon_r A V^2}{2(x_0 - x)^2}. \tag{10.11}$$

The movable electrode is suspended by springs with stiffness as shown in Figures 10.1 and 10.2. When the voltage is applied across the electrodes, the electrostatic attractive force F_E is induced. The induced attractive force leads to a decrease in the spacing, thereby stretching the spring. This results in an increase of the restoring spring force F_R, which counteracts the electrostatic force. According to Hook's law, restoring spring force F_R is presented by the following equation.

$$F_R = kx = F_E, \tag{10.12}$$

where k is the spring constant and x is the displacement of movable electrode structure. Therefore, the displacement x of the movable electrode structure can be expressed by

$$x = \frac{F_E}{k}. \tag{10.13}$$

10.3 DESIGN, MODELING, AND SIMULATION

10.3.1 RESONATOR STRUCTURE

The resonator structure with movable electrodes is shown in Figure 10.3. It basically consists of a silicon resonant body, movable electrodes, and stoppers. The resonant body is supported by two supporting beams on the sides and is placed between two electrodes (driving electrode and sensing electrode) with a narrow capacitive gap. Each electrode is supported by two springs and two beams, and can be electrostatically actuated. The stoppers are formed on the substrate, which make a narrow gap between the resonant body and movable electrodes. The parameters of the resonator are summarized in Table 10.1.

10.3.2 WORKING PRINCIPLE

The resonator is excited in its horizontal width extensional mode at the resonant frequency as the simulation result shows in Figure 10.4b, and the operation is described as follows. The direct current (DC) polarization voltage V_{DC} is applied to the driving/sensing movable electrodes while the resonant body is connected to ground. Thus, the same values of electrostatic forces ($F_1 = -F_2$, as shown in Figure 10.4a) are generated on both sides of the resonant body. When V_{DC} increases, the movable electrodes will move toward the resonant body due to the electrostatic force between the movable electrodes and the resonant body. The electrodes will be attracted to the resonant body and they touch the stoppers when V_{DC} is increased; thus, the capacitive gaps can be reduced.

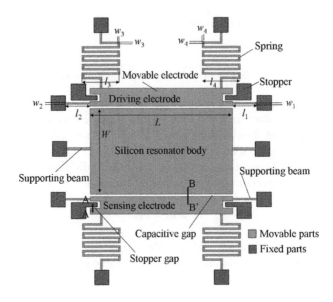

FIGURE 10.3 Schematic of silicon resonator structure with movable electrodes.

Figure 10.4a shows the schematic of the actuation obtained by finite element model (FEM) simulation using COMSOL software. Both movable electrodes move toward the resonant body when a DC voltage is applied. In this design, the reduced gap size (g_{reduced}) of the silicon resonator with the movable electrodes can be expected as given by

$$g_{\text{reduced}} = g_{B-B}{}' - g_{A-A}{}',$$
(10.14)

where $g_{B-B}{}'$ is the capacitive gap width and $g_{A-A}{}'$ is the stopper gap width, as shown in Figure 10.3.

The capacitive gap width is designed to be larger than the stopper gap width to make smaller gaps between the resonant body and electrodes when electrodes are actuated. In addition, when an AC voltage (V_{AC}) is applied to the driving electrode, the resulting electrostatic force causes a bulk acoustic wave in the resonant body. The resonator is electrostatically exited and vibrated at a horizontal width extensional mode for the rectangular plate by AC voltage in addition to DC voltage. The output voltage of the resonator results from changes in the capacitive gap on the sensing electrode.

10.3.3 DESIGN OF MOVABLE ELECTRODE STRUCTURES

The movable electrode structure is the critical component of this device. It is designed as a bar structure with the same length to the resonant body and supported by two beams and two springs. Both of the beams and springs are designed to be symmetric structures. The two beams and two springs are arranged in parallel with stiffness constants of k_{11}, k_{22}, k_{21}, and k_{22}, respectively, as shown in Figure 10.5a. The parameters of the movable electrode structure are summarized in Table 10.1.

TABLE 10.1
Summarized Parameters of the Silicon Resonators with and without Movable Electrodes

	With Movable Electrodes (Resonator 1)	Without Movable Electrodes (Resonator 2)
Parameters		
Length of resonant body, L	500 μm	500 μm
Width of resonant body, W	440 μm	440 μm
Thickness of SOI device layer, t	5 μm	5 μm
Capacitive gap, g	100 nm	400 nm
Length of the supporting beams, l_1, l_2	100 μm	None
Width of the supporting beams, w_1, w_2	3 μm	None
Length of the springs, l_3, l_4	120 μm	None
Width of the springs, w_3, w_4	6 μm	None
Applied Conditions		
V_{DC}	25 V	25 V
V_{AC}	0 dBm	0 dBm
Pressure level of vacuum chamber	0.01 Pa	0.01 Pa
Theoretical Calculation		
Effective mass of resonant body, m_{eff}	1.28×10^{-9} Kg	1.28×10^{-9} Kg
Effective stiffness of resonant body, k_{eff}	4.71×10^{6} Nm^{-1}	4.71×10^{6} Nm^{-1}
Effective stiffness of springs, k_1 and k_2	13.2 Nm^{-1}	None
Effective stiffness of beams, k_3 and k_4	22.8 Nm^{-1}	None
The total stiffness constant, k_{tot}	72 Nm^{-1}	None
Movable electrode displacement at 25 V, D	385 nm	None
Resonance frequency, f_0	9.67 MHz	9.67 MHz
Simulation Results		
Resonance frequency, f_0	9.75 MHz	9.75 MHz
Movable electrode displacement at 25 V	402 nm	None
Measurement Results		
Measured frequency, f_0	9.652260 MHz	9.654576 MHz
Quality factor, Q	48,607	63,952
Insertion loss, IL	−50.908 dB	−71.938 dB
Motional resistance, R_m	0.5 kΩ	100 kΩ
Motional capacitance, C_m	6.50×10^{-16} F	2.54×10^{-18} F
Motional inductance, L_m	4.18×10^{-1} H	1.07×10^{2} H
Feed-through capacitance, C_f	2.21×10^{-13} F	5.53×10^{-14} F
Tuning slope frequency	−253.5 Hz/V	−37.3 Hz/V

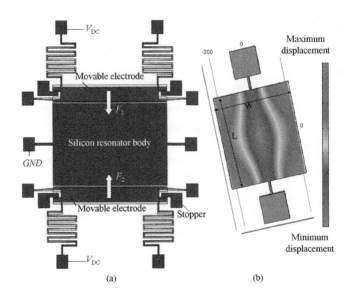

(a)   (b)

FIGURE 10.4 (a) FEM simulation showing displacement of the movable electrode. (b) Extensional mode of the resonant body at resonant frequency.

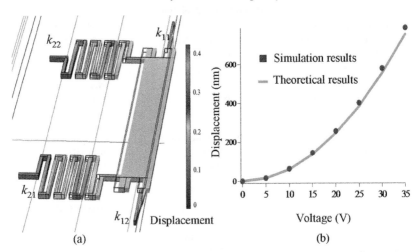

(a)   (b)

FIGURE 10.5 (a) FEM simulation result of the movable electrode structure. (b) Relationship of the displacement and applied voltage.

The electrode structures can move toward the resonant body due to electrostatic force, which is generated by the DC voltage as given by

$$F_E = \frac{1}{2} \varepsilon_0 \frac{Lt}{g^2} V_{DC}^2, \tag{10.15}$$

where F_E is the electrostatic force; V_{DC} is the polarization voltage; L and t are the length and the thickness of the resonant body, respectively; and g is the capacitive gap between the electrode and the resonant body. The stiffness constant k_m of the springs and supporting beams of the movable electrode can be expressed as follows: [19, 23]

$$k_m = \frac{Et_m w_m^3}{nl_m^3}, \tag{10.16}$$

where E is the Young's modulus of silicon material ($E = 169$ GPa); n is the number of beams for springs and $n = 1$ for the supporting beam; and t_m, w_m, and l_m are the thickness, width, and length of the beams (m = 11, 12) or springs (m = 21, 22), respectively. The total stiffness constant k_{tot} of all support parts (two springs and two supporting beams) for the movable electrodes is given as

$$k_{tot} = k_{11} + k_{12} + k_{21} + k_{22} = 2(k_{11} + k_{21}). \tag{10.17}$$

Using the parameters shown in Table 10.1, and also using Equations 10.16 and 10.17, the total stiffness constant k_{tot} for the movable electrode structures is calculated to be 72 Nm^{-1}. The displacement x of the movable electrode structure can be calculated by combining Equation 10.15 and Equation 10.17, as follows:

$$x = \frac{F_E}{k_{tot}}. \tag{10.18}$$

The FEM simulation results are in agreement with the calculation results, as shown in Figure 10.5a and b. The electrode structure will move toward the resonant body around 400 nm when 25 V of V_{DC} is applied. The relationship of the displacement and the applied voltage V_{DC} is plotted in Figure 6.5b. Lager displacement of movable electrode structures and lower applied voltage can be obtained by smaller springs and supporting beams. However, too soft springs and beams dramatically increase the probability of stiction, resulting in yield loss; therefore, it may cause unbalance for the movable electrode. Additionally, too soft springs and beams also possibly cause a thermomechanical noise and an ambient vibration noise at the low frequency region. In this regard, the parameters of springs and beams should be considered to obtain optimum performance of the displacement of the movable electrodes with the lowest polarization voltage.

10.4   FABRICATED RESULTS

Silicon resonator structures are formed by a combination of electron beam lithography (EBL) and photolithography, and deep reactive ion etching (DRIE) on the silicon on insulator (SOI) wafer, and then transferring to a low temperature co-fired ceramic (LTCC) substrate by using an anodic bonding technique. The handling silicon layer of the SOI wafer is etched out by plasma etching, and the buried SiO_2 layer is removed. The fabrication process of silicon resonators was presented in detail in previous chapters. The resonant structures with and without movable electrodes are successfully fabricated as shown in Figure 10.6a and c. The stopper gap width with A–A' cross section and the capacitive gap width with B–B' cross section of

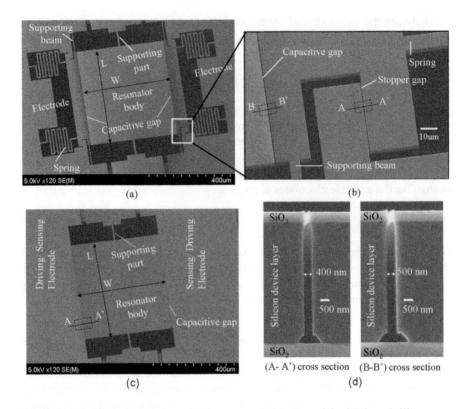

FIGURE 10.6 (a) Scanning electron microscope (SEM) view of the fabricated silicon resonator structure with movable electrode structures. (b) A close-up of a corner of the resonator
structure. (c) SEM photography of the fabricated silicon resonator without movable electrode
structures. (d) Profile shapes (A–A') and (B–B') cross section.

the structure with the movable electrodes are 400 nm and 500 nm, respectively, as
shown in Figure 10.6b and d. Therefore, the final gap widths can reach 100 nm in
this design, as discussed in Section 10.3. The silicon resonator structure without the
movable electrode with a 400 nm capacitive gap is fabricated as the A–A' cross section as shown in Figure 10.6d.

10.5   MEASUREMENT RESULTS

The resonance characteristics of silicon resonator structures with (resonator 1) and
without (resonator 2) movable electrodes is observed. The specifications are summarized in Table 10.1. As shown in Figure 10.7, the measurement results of the transmission S_{21} and phase response are indicated for resonator 1 with a length of 500 μm,
width of 440 μm, thickness of 5 μm, and reduced capacitive gap size of 100 nm. A
resonant peak, which is observed under measurement conditions V_{DC} of 30 V and
V_{AC} of 0 dBm, is found at 9.65 MHz with high Q factor of 50,000. Figure 10.7 also
presents a comparison of measured frequency characteristics in the vacuum chamber

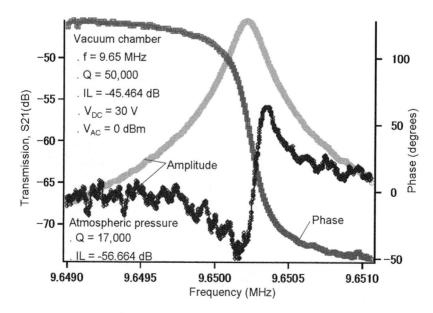

FIGURE 10.7 Transmission and phase response of resonator with movable electrode structures (resonator 1) and frequency response of resonator 1 both under vacuum chamber pressure and atmospheric pressure.

with pressures of 0.01 Pa and ambient atmosphere pressure of 10^5 Pa. The Q factor of 50,000 at 0.01 Pa is larger than that of 17,000 in ambient atmosphere due to absence of air viscous damping.

Although resonators 1 and 2 have been designed with the same dimensions and fabricated on the same SOI wafer, their resonance frequencies are a little different as we can see from Figure 10.8a (9.652260 MHz) and 10.8b (9.654576 MHz), respectively. This is due to fabrication tolerances such as scalloping, notching, and over-etching that make the effective spring constant and effective mass change slightly or the different measurement conditions such as pressure level and temperature. A comparison of the performance of the resonators 1 and 2 is shown in Table 10.1.

The resonant characteristics of resonators 1 and 2 with the different V_{DC} values are summarized in Table 10.2. It is shown that at a low V_{DC} of 5 V, the Q factor, calculated by the network analyzer based on its associated resonator peak, is not obtained for both resonators because the amplitude of peak resonant frequency is smaller than 3 dB. A small vibration signal with the resonant amplitude of 2 dB is observed with resonator 2, while that of resonator 1 is none. One of the possible reasons is due to the different capacitive gap widths. With a V_{DC} of 15 V, the resonant peaks are clearly observed and a high Q factor is achieved with both resonators. The final expected gap width of the silicon resonator with movable electrodes is around 365 nm at V_{DC} of 15 V, whereas the gap width of the silicon resonator without movable electrodes is 400 nm. The capacitive gap widths of the two resonators are not very different; therefore the frequency responses as well as other characteristics of both resonators are almost the same. The output signal of resonator 1 is 4 dB higher

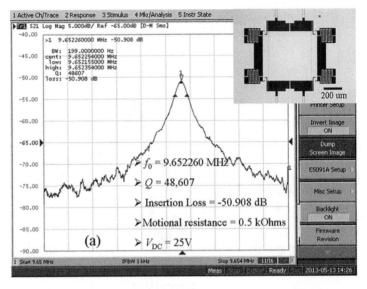

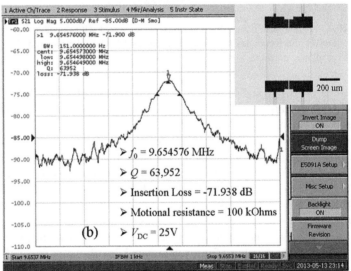

FIGURE 10.8   (a) Frequency response of resonator 1. (b) Frequency response of resonator 2.

than that of resonator 2, and there is a 2 dB difference in resonant amplitude between them. The frequency responses of resonators 1 and 2, which are performed under measurement conditions V_{DC} of 25 V and V_{AC} of 0 dBm, are shown in Figure 10.8. The measured transmission results in Figure 10.8 clearly show a 21 dB increase at resonance for the structure with the movable electrodes in comparison with the structure without the movable electrodes.

The calculation of the equivalent circuit model of resonator 1 based on measured data shows motional resistance $R_m = 0.5$ kΩ, motional capacitance $C_m = 6.50 \times 10^{-16}$ F,

TABLE 10.2

Summarized Resonant Characteristic of the Silicon Resonators with Different V_{DC}

	With Movable Electrodes (Resonator 1)	Without Movable Electrodes (Resonator 2)
$V_{DC}=5$ V	Final gap width, $g=485$ nm	Capacitive gap, $g=400$ nm
	Q factor, $Q=$Cannot be observed	Q factor, $Q=$Cannot be observed
	Insertion loss, IL $=-75.874$	Insertion loss, IL $=-75.163$
	Resonant amplitude, $A=0$ dB	Resonant amplitude, $A=2$ dB
$V_{DC}=15$ V	Final gap width, $g=365$ nm	Capacitive gap, $g=400$ nm
	Quality factor, $Q=37,579$	Quality factor, $Q=35,256$
	Insertion loss, IL $=-67.074$ dB	Insertion loss, IL $=-71.248$ dB
	Resonant amplitude, $A=11$ dB	Resonant amplitude, $A=9$ dB
$V_{DC}=25$	Final gap width, $g=100$ nm	Capacitive gap, $g=400$ nm
	Quality factor, $Q=48,607$	Quality factor, $Q=63.952$
	Insertion loss, IL $=-50.908$ dB	Insertion loss, IL $=-71.938$
	Resonant amplitude, $A=26$ dB	Resonant amplitude, $A=18$ dB

and motional inductance $L_m=4.18\times10^{-1}$ H in the reduced capacitive gap width of 100 nm, whereas that of resonator 2 shows motional resistance $R_m=100$ kΩ, motional capacitance $C_m=2.54\times10^{-18}$ F, and motional inductance $L_m=1.07\times10^2$ H in the capacitive gap width of 400 nm. The motional resistance value of resonator 1 is 200 times smaller than that of resonator 2 because of the smaller capacitive gap. However, the quality factor decreases from 64,000 to 49,000 in comparison of resonators 1 and 2.

The drop in the measured Q is found to be attributed to the various loading effects such as any series resistance in the resonator path, including termination resistors, as well as other resistances in the silicon structure of the resonator [10, 11]. All the loading effects can be collected in the equivalent resistor R_{loaded}, connected in series with the resonator. The equivalent circuit model for the silicon resonator with loading effects is presented in Figure 10.9b. The definitions of the unloaded quality factor and loaded quality factor are presented in [7, 10, 24] and as shown next:

$$Q_{unloaded} = \frac{2\pi f_0 L_m}{R_m}, \tag{10.19}$$

$$Q_{loaded} = \frac{2\pi f_0 L_m}{R_m + R_{loaded}}. \tag{10.20}$$

$Q_{unloaded}$ represents for the intrinsic properties, while Q_{loaded} shows the measured Q factor including the loading effect of R_{loaded}. The relationship between $Q_{unloaded}$ and Q_{loaded} can be found by combining Equations 10.19 and 10.20:

$$Q_{loaded} = Q_{unloaded} \frac{R_m}{R_m + R_{loaded}}. \tag{10.21}$$

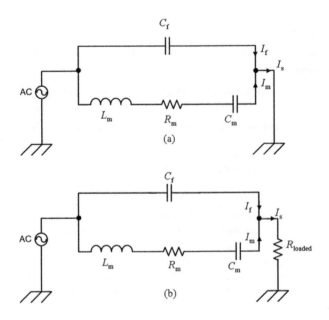

FIGURE 10.9   (a) Equivalent circuit model of the capacitive silicon resonator. (b) Equivalent circuit model of the capacitive silicon resonator with loading effects.

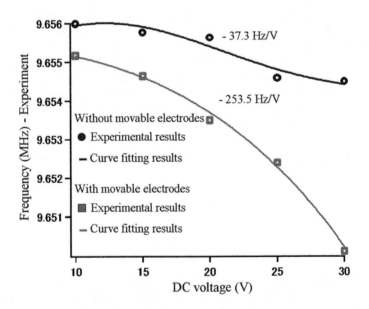

FIGURE 10.10   The electrostatic tuning characteristics of resonator 1 and resonator 2 in the theoretical calculation and experiment.

As discussed earlier, when the capacitive gap width reduces from 400 nm (resonator 2) to 100 nm (resonator 1), the motional resistance value is dramatically decreased 200 times from 100 kΩ to 0.5 kΩ. This causes the drop in the measured Q_{loaded} value, as presented in Equation 10.21. Therefore, the quality factor of the structure with movable electrodes is smaller than that of structure without movable electrodes.

The resonant peak shifts to lower frequency if the polarization voltage is increased due to the effect of electrical stiffness. The tuning frequency can be clearly explained from the following equation:

$$
f = \frac{1}{2\pi}\sqrt{\frac{k_{\text{eff}}}{m_{\text{eff}}}} = \frac{1}{2\pi}\sqrt{\frac{k_{\text{m}} + k_{\text{e}}}{m_{\text{eff}}}} = \frac{1}{2\pi}\sqrt{\frac{k_{\text{m}} - \dfrac{V_{\text{DC}}^2 \varepsilon_0 L t}{2g^3}}{m_{\text{eff}}}}, \tag{10.22}
$$

where k_{m} and k_{e} are mechanical and electrical spring constants, respectively. Figure 10.10 shows comparisons of the tuning slope between the structures with and without the movable electrodes. The measured electrostatic tuning characteristic for resonator 2 with 400 nm capacitive gap is around –37.0 Hz/V, while that of resonator 1 is around –253.0 Hz/V. The tuning characteristic of resonator 1 is 7 times that of resonator 2.

10.6 SUMMARY

The capacitive silicon resonator with movable electrode structures can reduce the motional resistance for lower insertion loss and also increase the tuning frequency range to compensate for the temperature drifts of the silicon resonators. Frequency characteristics of the silicon resonator of resonant frequency 9.65 MHz with a length of 500 μm, width of 440 μm, and thickness of 5 μm are evaluated, and a high Q factor of 49,000 is achieved at a polarization voltage of 25 V. The measurement results have shown that the motional resistance is reduced by 200 times, the output signal (insertion loss) is increased by 21 dB, and the tuning characteristic of the frequency is also increased by 7 times than that without movable electrode structures.

REFERENCES

1. Nguyen, C.T.C., MEMS technology for timing and frequency control, *IEEE Transactions on Ultrasonics, Ferroelectrics, and Frequency Control*, **54**, 251–270, 2007.
2. Akgul, M., Kim, B., Hung, L.W., Lin, Y., Li, W.C., Huang, W.L., Gurin, I., Borna, A., Nguyen, C.T.C., Oscillator far from carrier phase noise reduction via nano scale gap tuning of micromechanical resonator, *The 15th International Conference on Solid State Sensors, Actuators and Microsystems*, Denver, CO, 798–801, 2009.
3. Lee, S., Nguyen, C.T.C., Mechanical coupled micromechanical resonator arrays for improved phase noise, *Proceedings of IEEE International Conference on Ultrasonics, Ferroelectrics and Frequency Control*, 280–286, 2004.
4. Qishu, Q., Pourkamali, S., Ayazi, F., Capacitively coupled VHF silicon bulk acoustic wave filters, *IEEE Ultrasonics Symposium*, 1649–1652, 2007.

5. Demirci, M.U., Nguyen, C.T.-C., Mechanically corner-coupled square microresonator array for reduced series motional resistance, *Journal of Microelectromechanical Systems*, **15**, 1419–1436, 2006.
6. Bannon, F.D., Clark, J.R., Nguyen, C.T.-C., High-Q, HF micromechanical filters, *Journal of Solid-State Circuits*, **35**, 512–526, 2000.
7. van Beek, J.T.M.V., Puers, R., A review of MEMS oscillators for frequency reference and timing applications, *Journal of Micromechanics and Microengineering*, **22**, 013001, 2012.
8. Weinstein, D., Bhave, S.A., Piezoresistive sensing of a dielectrically actuated silicon bar resonator, *Solid-State Sensors, Actuators, and Microsystems Workshop*, 368–371, 2008.
9. Lin, Y.W., Li, S.S., Xie, Y., Ren, Z., Nguyen, C.T.-C., Vibrating micromechanical resonators with solid dielectric capacitive transducer gaps, *Frequency Control Symposium and Exposition*, 128–134, 2005.
10. Pourkamali, S., Ho, G.K., Ayazi, F., Low-impedance VHF and UHF capacitive silicon bulk acoustic wave resonators – Part I: Concept and fabrication, *IEEE Transactions on Electron Devices*, **54**, 2017–2023, 2007.
11. Pourkamali, S., Ho, G.K., Ayazi, F., Low impedance VHF and UHF capacitive silicon bulk acoustic wave resonators – Part II: Measurement and characterization. *IEEE Transactions on Electron Devices*, **54**, 2024–2030, 2007.
12. Ho, G.K., Sundaresan, K., Pourkamali, S., Ayazi, F., Micromechanical IBARs: Tunable high Q resonator for temperature compensated reference oscillators, *Journal of Microelectromechanical Systems*, **19**, 503–515, 2010.
13. Hsu, W.T., Clark, J.R., Nguyen, C.T.-C., A sub-micron capacitive gap process for multiple metal electrode lateral micromechanical resonators, *14th IEEE International Conference on Micro Electro Mechanical Systems*, 392–252, 2001.
14. Shao, C.Y., Kawai, Y., Esashi, M., Ono, T., Electrostatic actuator probe with curved electrodes for time of flight scanning force microscopy, *The Review of Scientific Instruments*, **81**, 083702–083706, 2010.
15. Barnes, S.M., Miller, S.L., Rodgers, M.S., Bitsie, F., Torsional ratcheting actuating systems, *Technical Proceedings of the 2000 International Conference on Modeling and Simulation of Microsystems*, San Diego, CA, 273–276, 2000.
16. Lee, Y., Toda, M., Esashi, M., Ono, T., Micro wishbone interferometer for miniature FTIR spectrometer, *IEEJ Transactions on Sensors and Micromachines*, **130**, 333–334, 2010.
17. Lee, Y.M., Toda, M., Esashi, M., Ono, T., Micro wishbone interferometer for flourier transform infrared spectrometry, *Journal of Micromechanics and Microengineering*, **21**, 065039, 2011.
18. Tanner, D.M., Owen, A.C., Rodriguez, F., Resonator frequency method for monitoring MEMS fabrication, *SPIE's Proceeding of Reliability, Testing and Characterization of MEMS/MOEMS*, **4980**, 220–228, 2003.
19. Liu, X., Kim, K., Sun, Y., A MEMS Stage for 3-axis nanopositioning, *Journal of Micromechanics and Microengineering*, **17**, 1796–1802, 2007.
20. Senturia, S.D., *Microsystem Design*, Kluwer Academic Publishers, Boston, 2001.
21. Muldavin, J.B., Rebeiz, G.M., Inline capacitive and DC contact MEMS shunt switches, *IEEE Microwave and Wireless Components Letters*, **11**, 334–336, 2001.
22. Hsu, W.T., Vibrating, R.F., MEMS for timing and frequency references, *Microwave Symposium Digest*, 672–675, 2006.
23. Tsay, J., Su, L.Q., Sung, C.K., Design of a linear micro-feeding system featuring bistable mechanisms, *Journal of Micromechanics and Microengineering*, **15**, 63–70, 2005.
24. Ho, G.K., Perng, J.K., Ayazi, F., Micromechanical IBARs: Modeling and process compensation, *Journal of Microelectromechanical Systems*, **19**, 516–525, 2010.

11 Capacitive Silicon Resonators with Piezoresistive Heat Engines

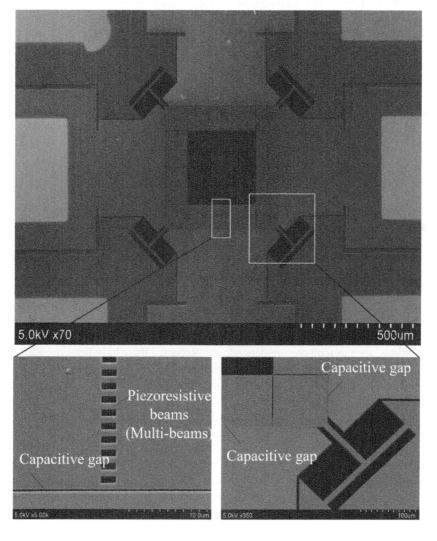

11.1 INTRODUCTION

Micromechanical resonators have been employed for a variety of applications [1–9]. Mass sensing with resonating ultra-thin silicon beams has been demonstrated [1–2], where the resonant frequency of an oscillating thin cantilever beam monitors a loaded mass. Timing reference applications were studied and presented in many works [3–6], where resonator structures can be used to generate clock signals in electronic systems. Microresonators are also employed for filtering applications [7–9], which can be present in radio frequency transmitter and receiver modules nowadays. To actuate and sense the motion of resonators, many transduction mechanisms, including piezoelectric [7–10], capacitive [5–6, 11] and piezoresistive [12–13], have been investigated. All the above methods possess both advantages and disadvantages. The capacitive technique, typically employed for sensing and timing references, is based on the measurement of the change in the capacitance between a sensing electrode and the resonant body. Capacitive resonators can show high stability and low phase noise; however, their drawbacks are large motional resistances and high insertion losses that make it difficult for them to meet oscillation conditions. Methods for lowering the motional resistance [14–21] utilize reducing the capacitive gap width [14–16], and increasing the overlap area of the capacitance, the quality factor Q, and the polarization voltage V_{DC}. Unfortunately, each of these methods has some drawbacks. Decreasing the capacitive gap width is very effective in lowering the motional resistance; however, the fabrication of a nanogap is very difficult, and the applicable maximum polarization voltage decreases due to the pull-in phenomenon. Smaller capacitive gaps result in lower pull-in voltages that will easily cause short-circuit situations. An increase in the electrode areas being utilized such as resonator arrays [4, 17] and mechanical coupling [18], can reduce the motional resistance, but the frequency response and Q factor of the resonators suffer from the mismatches in the resonant frequencies.

To increase the Q factor, some of the methods to reduce the energy losses, including external and internal losses, have been reported, but there are material and structural limitations [5, 19–21]. In addition, piezoresistive sensing techniques for micromechanical resonators have been studied [22, 23]. Piezoresistive sensing with thermal actuation shows a high quality factor and low insertion loss; however, it faces large power consumption. Besides, this technique is not appropriate for the construction of high-sensitivity products due to strong effects from environmental conditions, such as temperature dependence. Recently, there have been some efforts to improve piezoresistive transduction sensitivity [24, 25]. Self-sustained oscillators without any amplifying circuitry can be achieved by piezoresistive heat engines, which are based on thermodynamic cycles. Their operation principles are as follows: due to the higher resistance of narrow piezoresistive beams over other parts of the resonator structure, the piezoresistive beams are heated up when a direct current (DC) voltage is applied to the piezoresistive elements. This results in the expansion of the piezoresistive beams, which causes the resistance to increase due to the piezoresistive effect. As a result of this resistance increase, the current passing through the piezoresistive beams decreases, which also decreases the temperature of the piezoresistive beam. Thus, the piezoresistive beams are compressed and the resistance decreases due to the piezoresistive effect, resulting in an increase in the current. From this, a thermomechanical actuation power is generated by the described cycle in the piezoresistive elements. Therefore, self-oscillation can be achieved [24].

Achieving all the aforementioned advantages in different transduction mechanisms on a single resonator is highly expected. Combined capacitive and piezoelectric transductions for high-performance silicon micromechanical resonators have been developed [26]. Integrating capacitive actuation and piezoresistive sensing in micromechanical resonators has been demonstrated [27, 28].

In this chapter, capacitive silicon resonators with piezoresistive heat engines are proposed and examined. Piezoresistive thermal actuators are used for the excitation of the vibration to enhance the driving force. Two designs of capacitive resonators, which consist of single and multiple piezoresistive beams, are demonstrated. The fabricated devices are evaluated and compared with each other in cases with and without the piezoresistive effect.

11.2   DEVICE DESCRIPTION

11.2.1   Device Structure and Working Principle

The basic components of the proposed device's structure are schematically shown in Figure 11.1. It consists of the resonant body, supporting beams, electrodes, piezoresistive beams, and capacitive gaps. The resonant body is a square frame structure that is fixed at four corners of the square plate via the supporting beams. The resonant body is divided into many parts that are connected to others using small piezoresistive beams. Four electrodes create narrow capacitive gaps with the resonant body. The design parameters of the resonator are shown in Table 11.1. All structures are made of p-type, low-resistivity, single-crystal silicon of 0.02 Ω cm.

Two kinds of resonator designs are presented in this work. A single piezoresistive beam at the connection areas (connecting two parts of the resonant body) is used in device 1 (Figure 11.1a). The resonant body of device 1 is split into four parts that are connected to each other by the single piezoresistive beam above. This beam is placed at the center of the resonant body, as illustrated in Figure 11.1. In device 2, multiple piezoresistive beams (10 beams) have been employed, which are located near the edges of the resonant body to enhance the vibration amplitude. Its resonant body is divided into the 12 elements, as shown in Figure 11.1b.

The operation principle of these devices is as follows: The resonator structure works in a capacitive mode, in which the output voltage results from the changes in the capacitive gap on the sensing electrode. A current controlled by a voltage source V_b is applied through the resonant body. Due to the higher electrical resistance of the narrow actuator beams than that of the other parts of the body, Joule heating mostly occurs at the actuator beams. The vibration of the resonant body is mainly caused by capacitive transduction in addition to the piezoresistive engine using the piezoresistive beams. As previously described, the temperature modulation (thermodynamic cycles consist of heating and cooling cycles) in the piezoresistive beam causes thermomechanical force. Thus, the thermodynamic cycles (thermal actuators) are possibly used for an efficient excitation of vibration with the driving force enhancement.

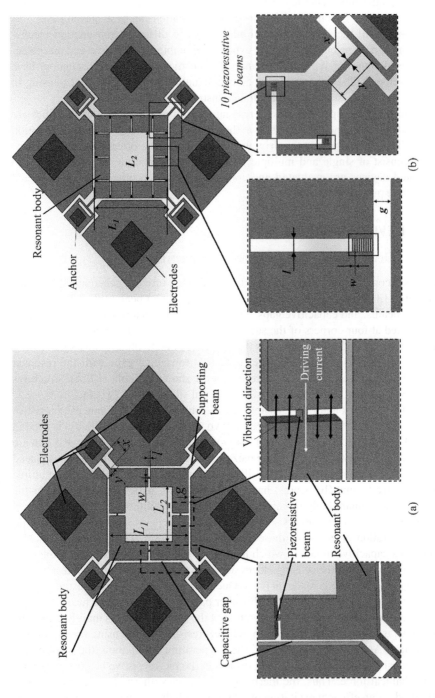

FIGURE 11.1 Capacitive silicon resonator with piezoresistive heat engines. (a) Device 1. (b) Device 2.

TABLE 11.1

Summary of Parameters of Devices 1 and 2

Device Parameters	Device 1 (Single Piezoresistive Beam)	Device 2 (Multiple Piezoresistive Beams)
Type of semiconductor	p-type	p-type
Electrical resistivity	0.02 Ω cm	0.02 Ω cm
Outer length/width of resonator body, L_1 (μm)	500	500
Inner length/width of resonator body, L_2 (μm)	400	400
Thickness of device layer, t (μm)	7	7
Capacitive gap, g (nm)	250	250
Length of supporting beam, x (μm)	100	100
Width of supporting beam, y (μm)	10	10
Length of piezoresistive beam, l (μm)	3	3
Width of piezoresistive beam, w (μm)	0.8	0.8
Number of piezoresistive beams	Single	Multiple
Resonant body is divided into	4	12

The resonant frequency f_0 of the fundamental mode can be calculated using the following equation:

$$f_0 = \frac{1}{2\Pi}\sqrt{\frac{k_{\text{eff}}}{m_{\text{eff}}}}, \tag{11.1}$$

where k_{eff} and m_{eff} are the effective spring constant and mass, respectively.

The motional resistance R_{m} represents the mechanical loss of the vibration and can be extracted from the insertion loss as follows [11]:

$$R_{\text{m}} = 50\left(10^{\frac{\text{IL}_{\text{dB}}}{20}} - 1\right), \tag{11.2}$$

where IL_{dB} is the insertion loss of the transmission and its unit is in decibels (dB).

A thermal response time τ and frequency response f_{T} of the piezoresistive beams can be estimated using its thermal resistance R_{T} and capacitance C_{T}:

$$R_{\text{T}} = \frac{l}{kwt}, \tag{11.3}$$

$$C_{\text{T}} = \rho C_{\text{mass}}lwt, \tag{11.4}$$

$$\tau = R_{\text{T}}C_{\text{T}}, \tag{11.5}$$

$$f_{\text{T}} = \frac{1}{2\pi\tau}, \tag{11.6}$$

where l, w, and t are the length, width, and thickness of the piezoresistive beam, respectively. k, ρ, and C_{mass} are the thermal conductivity, density, and specific heat of the silicon material, respectively.

A theoretical prediction of the resonant frequency of the proposal resonators is difficult due to the complex structures. To solve this problem, the finite element method (FEM) is employed for the vibration mode (resonant frequency and vibration shape) and temperature distribution, as presented in the following section.

11.2.2 Finite Element Method (FEM) Simulation

The FEM simulation results are shown in Figure 11.2. The vibration mode (Figure 11.2a and b) and temperature distribution (Figure 11.2c and d) in the proposed devices are depicted. Resonators are electro-statically excited and vibrated at the resonant frequency of the extensional mode. The highest temperature is shown at the piezoresisitve beams. Devices 1 and 2 are vibrated at resonant frequencies of 3.16 MHz and 1.29 MHz, respectively, as the vibration mode is illustrated in Figure 11.2a and b. The maximum displacement is located at the center of the square resonant edges where the piezoresistive beam is placed.

The simulated temperature distribution in the designed resonators with the bias current (voltage source, V_b) running through the resonant body via two anchors is shown in Figure 11.2c and d. The highest recorded temperature of 293°C is at the piezoresistive beam when the voltage source $V_b = 7$ V is supplied.

11.3 EXPERIMENTS AND DISCUSSIONS

11.3.1 Fabrication

A combination of electron beam lithography (EBL), photolithography, and deep reactive ion etching (DRIE) is proposed to create the nanotrenches and resonator structures. The formed silicon resonator structures on the silicon on insulator (SOI) wafer are transferred to a glass substrate by anodic bonding in order to reduce the parasitic capacitances from the handling silicon layer. The fabrication process was presented in Chapter 9. Before device fabrication, an investigation of the nanotrenches is performed and the experimental result is shown in Figure 11.3. Nanotrenches from 200 nm to 500 nm (Figure 11.3a) with smooth surfaces (very low scallop, as shown in Figure 11.3b) and vertical shapes can be achieved. The fabricated devices 1 and 2 are shown in the Figure 11.4a and b, respectively.

11.3.2 Measurement Setup

The experimental setup used for the evaluation of the resonant characterization of the devices is illustrated in Figure 11.5. The typical displacements for the capacitive silicon resonators are at the nanometer scale and their vibrations are seriously damped at an ambient pressure; therefore, the fabricated devices are evaluated in a vacuum chamber at a low pressure of 0.01 Pa. The resonator electrodes are connected to a circuit, sources, and network analyzer outside the chamber by using

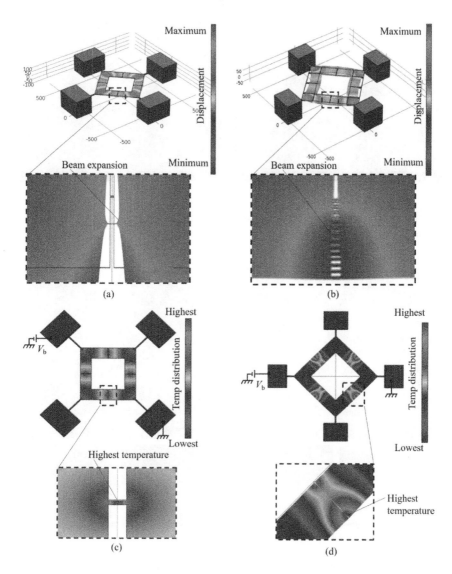

FIGURE 11.2 FEM simulation results. (a) Vibration mode of device 1. (b) Vibration mode of device 2. (c) Temperature distribution in device 1. (d) Temperature distribution in device 2.

the feed-through electrical connectors. The frequency response of the fabricated resonators is measured by a vector network analyzer (E5071B ENA Series, Agilent Technologies) with a frequency range from 300 kHz to 8.5 GHz.

There are two kinds of measurement setups. The first measurement setup surveys the capacitive effect. In this setup, a DC voltage V_{DC} is applied to the driving and sensing electrodes against the grounded resonator body through a 100 kΩ resistor, which is decoupled from the RF output of the network analyzer using a 100 nF capacitor. The vibration characteristics are obtained by the capacitive detection between

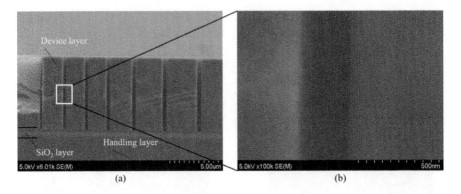

FIGURE 11.3 Testing results of nanotrenches. (a) Nanotrenches from 200 nm to 500 nm. (b) Close-up image of the etched surface.

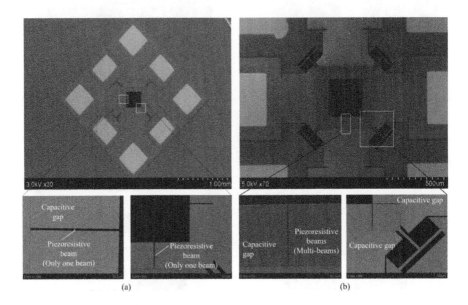

FIGURE 11.4 Fabricated devices. (a) Capacitive silicon resonator with a single piezoresistive beam. (b) Capacitive silicon resonator with multiple piezoresistive beams.

the sensing electrode and resonant body. Small changes in the capacitive gap generate a voltage on the RF input of the network analyzer. The second measurement setup is used for the assessment of the piezoresistive effect on the capacitive silicon resonators. The measurement setup is similar to that described earlier; however, an additional bias voltage V_b is applied to the piezoresistive elements to generate the current through the resonant body, as shown in Figure 11.5. In this setup, piezoresistive beams are used as thermal actuators to enhance the driving force, as mentioned in an earlier section. In other words, the frequency response of the fabricated devices

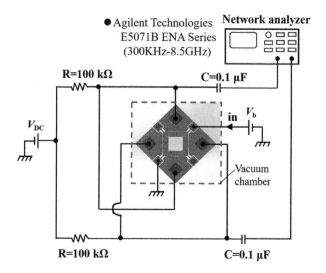

FIGURE 11.5 Measurement setup.

is evaluated in the cases without (capacitive effect) and with (piezoresistive effect) the bias voltage V_b through the resonant body.

11.3.3 MEASUREMENT RESULTS

The specifications of the fabricated devices are summarized in Table 11.2. A resonant peak for device 1 is observed under the measurement conditions V_{DC} of 20 V and source of power V_{AC} of 0 dBm, and found at 3.166 MHz with a quality factor Q of 3000. The measured resonant frequency is in good agreement with the FEM simulation results (Figure 11.2a and b). Figure 11.6a shows the frequency response

TABLE 11.2

Summary of Measurement Results of Devices 1 and 2

Evaluated Results	Device 1 (Single Piezoresistive Beam)	Device 2 (Multiple Piezoresistive Beams)
Resonant frequency	3.166 MHz	1.333 MHz
Quality factor (Q)	3000	2000
Insertion loss		
$V_{bias} = 0$ V	−50 dB	−68 dB
$V_{bias} = 7$ V	−40 dB	−48 dB
Enhanced	**10 dB**	**20 dB**
Motional resistance		
$V_{bias} = 0$ V	15.7 kΩ	125.5 kΩ
$V_{bias} = 7$ V	4.9 kΩ	12.5 kΩ
Reduced	**70%**	**90%**

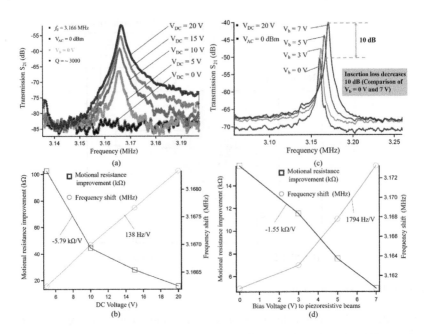

FIGURE 11.6   Measurement results of device 1. (a) Capacitive effect. (b) Motional resistance improvement and resonant frequency shift under polarization voltage V_{DC}. (c) Piezoresistive heat engine effect. (d) Motional resistance improvement and resonant frequency shift under thermal actuator V_b.

of device 1 without connecting the bias voltage V_b (capacitive effect only) under measurement conditions of $V_{AC} = 0$ V dBm, $V_b = 0$ V, pressures of 0.01 Pa, and various values of V_{DC}. When $V_{DC} = 0$ V and $V_{AC} = 0$ dBm ($V_{peak\ to\ peak} = 0.6$ V), no resonant peak can be found because there is no electrostatic actuation on the sensing gaps. By increasing the polarization voltage V_{DC} from 5 V to 20 V, the amplitude transmission S_{21} is boosted from −66 dB to −50 dB, as shown in Figure 11.6a. A motional resistance improvement of −5.79 kΩV^{-1} was achieved by tuning V_{DC} (Figure 11.6b). The vibration amplitude and motional resistance can be improved by raising the polarization voltage V_{DC}; however, it would be limited because the pull-in voltage appears due to the small capacitive gap. In addition, a high voltage is not available on the complementary metal–oxide semiconductor (CMOS) chip, which makes resonators difficult for large-scale integration (LSI). Figure 11.6c shows the frequency response when the bias voltage V_b is supplied to the piezoresistive elements (engine). Its output signal (insertion loss) is enhanced by 10 dB (from −50 dB to −40 dB) in this case. The Q factor (Q ~3000) of device 1 is almost unchanged when applying the bias voltage V_b (from 0 V to 7 V) to the piezoresistive beams. Figure 11.6b and d show the measured frequency shifts of device 1 caused by the polarization voltage V_{DC} and thermal actuators V_b (piezoresistive heat engines), respectively. The measured electrostatic tuning characteristic is 138 HzV^{-1}, while that caused by the piezoresistive heat engines is 1794 HzV^{-1}. The principle mechanism relies on the resonant frequency shift caused by the physical change such as a loaded mass, an applied force, or temperature environment [1, 2, 15, 26]. Typically, the resonant peak shifts to a slightly lower frequency if the polarization voltage V_{DC} is increased because of the effect of the electrical stiffness, as mentioned in our previous works [6, 16]. However, the resonant frequency of device 1 is shifted to the right side when increasing the polarization voltage V_{DC} (Figure 11.6a). The reason may come from the nonlinear hardening resonance (hardening effect) [29] or nonlinear damping effect (large vibration amplitude) [30]. A motional resistance improvement of -1.55 kΩV^{-1} has been achieved for this device with thermal actuators V_b (Figure 11.6d).

To enhance the piezoresistive electrothermal force, device 2 with multiple piezoresistive beams, which are placed near the edges of the resonant body, is examined. The resonant frequency of device 2 is observed at 1.333 MHz, which is smaller than that of device 1. The resonant body of device 2 is divided into 12 parts, while that of the device 1 is in 4 parts. Dividing the resonant body into more sections and embedding more piezoresistive beams within the resonant body can cause a lower stiffness because of their smaller widths. Thus, it results in the lower resonant frequency.

The FEM simulation result is in agreement with this result, as shown in Table 11.1. A 20 dB improvement (from -68 dB to -48 dB) in the insertion loss is achieved when a 7 V bias voltage V_b is supplied (Figure 11.7). Nevertheless, the peak shape

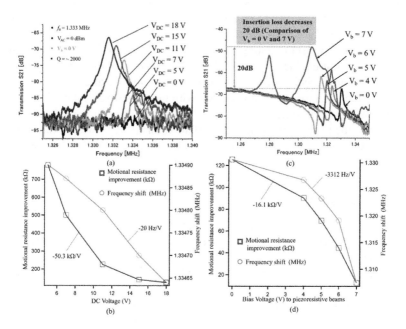

FIGURE 11.7 Measurement results of device 2. (a) Capacitive effect. (b) Motional resistance improvement and resonant frequency shift under polarization voltage V_{DC}. (c) Piezoresistive heat engine effect. (d) Motional resistance improvement and resonant frequency shift under thermal actuator V_b.

becomes complex. There are some possible reasons for this behavior. A thermo-mechanical coupling with a shifted phase may be one of the reasonable causes. A calculation of the thermal cut-off frequency (frequency response of the piezoresistive beams) based on Equation 11.6 has been performed and its value is 1.62 MHz. The resonant frequency of device 2 is below the thermal cut-off frequency. Thus, there is no phase shift between the stress in the piezoresistive beams and their amplitude in device 2, whereas for device 1 there will be a 90° phase lag in the temperature fluctuations with respect to the displacement. This is probably why the two designs show different behaviors under the high bias voltage supply to the piezoresistive beams. The other possible reasons may come from mechanical saturation that leads to an unstable region [25] or a thermoelastic dissipation [31] that causes a high energy loss. The Q factor of approximately 2000 is stable as a bias voltage V_b in the range 0 V to 6 V. However, a drop in the Q factor (from 3000 to 260) is found at the bias voltage V_b of 7 V. The possible reasons for this are mentioned in the earlier discussion. The motional resistance of both devices can be extracted from the insertion loss, as mentioned in Equation 11.2. By tuning the bias voltage V_b from 0 V to 7 V, the motional resistance of devices 1 and 2 is tuned from 15.7 kΩ to 4.9 kΩ and 125.5 kΩ to 12.5 kΩ, respectively. Thus, reductions of 70% and 90% can be achieved for devices 1 and 2, respectively.

From Figure 11.7b and d, it can be seen that the measured resonant frequency of device 2 also shifts when increasing the polarization voltage V_{DC} or thermal actuators V_b. A reduction in the resonant frequency in this device is attributed to the electrical spring softening known to occur in capacitively actuated resonators [6, 16]. The measured electrostatic tuning characteristic is −20 HzV^{-1}, while that caused by piezoresistive heat engines is 3312 HzV^{-1}. Thus, there is a 165 times difference in the tuning characteristic between the polarization voltage and thermal actuators. Therefore, the capacitive silicon resonator with piezoresistive heat engines can also enhance the tuning frequency range, which would be one of the methods for compensation of the temperature drifts in oscillation for timing device application. The motional resistance improvements of −50.3 kΩV^{-1} and −16.1 kΩV^{-1} under the V_{DC} and V_b are also shown in Figure 11.7b and d, respectively.

11.4   SUMMARY

In this chapter, we proposed and demonstrated that piezoresistive beams (piezoresistive heat engines) can enhance the performance of capacitive silicon resonators. Capacitive silicon resonators with single and multiple piezoresistive beams were fabricated and evaluated. Improvements in the insertion loss and reductions in the motional resistance were achieved. With the bias voltage $V_b = 7$ V supply, the insertion loss of the fabricated devices is enhanced by 10 dB and 20 dB for device 1 and device 2, respectively, in comparison with the capacitive silicon resonators without power to the piezoresistive heat engine. The motional resistance of devices 1 and 2 is reduced by 70% and 90%, respectively. Additionally, the tuning frequency characteristic with the piezoresistive effect is increased by 165 times over that of the structure with only the capacitive effect.

REFERENCES

1. Ono, T., Esashi, M., Mass sensing with resonating ultra-thin silicon beams detected by a double-beam laser Doppler vibrometer, *Measurement Science and Technology*, **15**, 1977–1981, 2004.
2. Kim, S.J., Ono, T., Esashi, M., Mass detection using capacitive resonant silicon resonator employing LC resonant circuit technique, *Review of Scientific Instrumentation*, **78**, 085103, 2007.
3. Nguyen, C.T.C., MEMS technology for timing and frequency control, *IEEE Transactions on Ultrasonics, Ferroelectrics, and Frequency Control*, **54**, 251–270, 2007.
4. Ayazi, F., MEMS for integrated timing and spectral processing, *IEEE Custom Integrated Circuits Conference*, Rome, Italy, 65–72, 2009.
5. van Beek, J.T.M., Puers, R., A review of MEMS oscillations for frequency reference and timing applications, *Journal of Micromechanics and Microengineering*, **22**, 013001, 2012.
6. Toan, N.V., Miyashita, H., Toda, M., Kawai, Y., Ono, T., Fabrication of an hermetically packaged silicon resonator on LTCC substrate, *Microsystem Technologies*, **19**, 1165–1175, 2013.
7. Piazza, G., Stephanou, P.J., Porter, J.M., Wijesundara, M.B.J., Pisano, A.P., Low motional resistance ring-shaped contour-mode aluminum nitride piezoelectric micromechanical resonators for UHF application, *18th IEEE International Conference on Micro Electro Mechanical Systems*, Miami Beach, FL, 20–23, 2005.
8. Sorenson, L., Fu, J.L., Ayazi, F., One-dimensional linear acoustic bandgap structures for performance enhancement of AlN-on-silicon micromechanical resonators, *16th International Conference on Solid State Sensors, Actuators and Microsystems*, Beijing, China, 918–921, 2011.
9. Nguyen, N., Johannesen, A., Hanke, U., Design of high Q thin film bulk acoustic resonator using dual-mode reflection, *IEEE International Ultrasonics Symposium*, Chicago, IL, 487–490, 2014.
10. Ali, A., Lee, J.E.-Y., Novel platform for resonant sensing in liquid with fully electrical interface based on an in-plane-mode piezoelectric-on-silicon resonator, *Procedia Engineering*, **120**, 1217–1220, 2015.
11. Ho, G.K., Sundaresan, K., Pourkamali, S., Ayazi, F., Low motional impedance highly tunable I2 resonators for temperature compensated reference oscillators, *18th IEEE International Conference on Micro Electro Mechanical Systems*, Miami Beach, FL, 116–120, 2005.
12. van Beek, J.T.M., Steeneken, P.G., Giesbers, B., A 10 MHz piezoresistive MEMS resonator with high Q, *IEEE International Frequency Control Symposium and Exposition*, Miami, FL, 475–480, 2006.
13. Rahafrooz, A., Pourkamali, S., Thermal-piezoresistive energy pumps in micromechanical resonator structures, *IEEE Transactions on Electron Devices*, **59**, 3587–3593, 2012.
14. Pourkamali, S., Ho, G.K., Ayazi, F., Low impedance VHF and UHF capacitive silicon bulk acoustic wave resonators – Part I: Concept and fabrication, *IEEE Transactions on Electron Devices*, **54**, 2017–2023, 2007.
15. Pourkamali, S., Ho, G.K., Ayazi, F., Low impedance VHF and UHF capacitive silicon bulk acoustic wave resonators – Part II: Measurement and characterization, *IEEE Transactions on Electron Devices*, **54**, 2024–2030, 2007.
16. Toan, N.V., Toda, M., Kawai, Y., Ono, T., A capacitive silicon resonator with a movable electrode structure for gap width reduction, *Journal of Micromechanics and Microengineering*, **24**, 025006, 2014.

17. Qishu, Q., Pourkamali, S., Ayazi, F., Capacitively coupled VHF silicon bulk acoustic wave filters, *IEEE Ultrasonics Symposium*, 1649–1652, 2007.
18. Toan, N.V., Shimazaki, T., Ono, T., Single and mechanically coupled capacitive silicon nanomechanical resonators, *Micro & Nano Letters*, **11**, 591–594, 2016.
19. Lee, J.E.Y., Yan, J., Seshia, A.A., Study of lateral mode SOI-MEMS resonators for reduced anchor loss, *Journal of Micromechanics and Microengineering*, **21**, 045011, 2011.
20. Toan, N.V., Kubota, T., Sekhar, H., Samukawa, S., Ono, T., Mechanical quality factor enhancement in silicon micromechanical resonator by low-damage process using neutral beam etching technology, *Journal of Micromechanics and Microengineering*, **24**, 085005, 2014.
21. Toan, N.V., Toda, M., Kawai, Y., Ono, T., A long bar type silicon resonator with a high quality factor, *IEEJ Transactions on Sensors and Micromachines*, **134**, 26–31, 2014.
22. van Beek, J.T.M., Steeneken, P.G., Giesbers, B., A 10 MHz piezoresistive MEMS resonator with high Q, *IEEE International Frequency Control Symposium and Exposition*, Miami, FL, 475–480, 2006.
23. Rahafrooz, A., Pourkamali, S., Thermal-piezoresistive energy pumps in micromechanical resonator structures, *IEEE Transactions on Electron Devices*, **59**, 3587–3593, 2012.
24. Steeneken, P.G., Le Phan, K., Goossens, M.J., Koops, G.E.J., Brom, G.J.A.M., van der Avoort, C., van Beek, J.T.M., Piezoresistive heat engine and refrigerator, *Nature Physics*, **7**, 354–359, 2011.
25. Ramezany, A., Mahdavi, M., Pourkamali, S., Nanoelectromechanical resonant narrow band amplifiers, *Microsystems & Nanoengineering*, **2**, 16004, 2016.
26. Samarao, A.K., Ayazi, F., Combined capacitive and piezoelectric transduction for high performance silicon microresonators, *24th IEEE International Conference on Micro Electro Mechanical Systems*, Cancun, Mexico, 169–172, 2011.
27. van Beek, J.T.M., Verheijden, G.J.A.M., Koops, G.E., Phan, K.L., Avoort, C.V.D., Scalable 1.1 GHz fundamental mode piezo-resistive silicon MEMS resonator, *IEEE International Electron Devices Meeting*, Washington, DC, 411–414, 2007.
28. Weinstein, D., Bhave, S.A., Piezoresistive sensing of a dielectrically actuated silicon bar resonator, *Solid-State Sensors, Actuators, and Microsystems Workshop*, Hilton Head Island, SC, 368–371, 2008.
29. Cho, H., Jeong, B., Yu, M.F., Vakakis, A.F., McFarland, D.M., Bergman, L.A., Nonlinear hardening and softening resonance in mechanical cantilever-nanotube systems originated from nanoscale geometric nonlinearities, *International Journal of Solids and Structures*, **49**, 2059–2065, 2012.
30. Inomata, N., Saito, K., Ono, T., Q factor enhancement of Si resonator by nonlinear damping, *Microsystem Technologies*, **23**, 1201–1205, 2017.
31. Chandorkar, S.A., Candler, R.N., Duwel, A., Melamud, R., Agarwal, M., Goodson, K.E., Kenny, T.W., Multimode thermoplastic dissipation, *Journal of Applied Physics*, **105**, 043505, 2009.

12 Conclusions

Various methods of enhancing the performance of capacitive silicon resonators were demonstrated in this book. The different novel fabrication technologies, including hermetic packaging based on low temperature co-fired ceramic (LTCC) substrate, deep reactive ion etching, neutral beam etching technology, and metal-assisted chemical etching, as well as design considerations, consisting of mechanically coupled, selective vibration of the high-order mode, movable electrode structures, and piezoresistive heat engines were investigated to increase the Q factor, decrease motional resistance and insertion loss, and widen the tuning frequency range. The main achievements and original contributions of this book can be summarized as follows:

- A wide range of micro/nano fabrication technologies was investigated to optimize the performance of micromechanical silicon resonators. Many challenges and difficulties were overcome, including:
  - A combination of electron beam lithography and standard photolithography was proposed to provide the possibility of fabricating micromechanical silicon resonators with narrow capacitive gap width and to be less time-consuming.
  - A thick SiO_2 was etched with nano gap width and vertical shape profile for the etching mask of the silicon device layer.
  - A vertical shape profile with small scallops of around 10 nm and sub-micro capacitive gap width with deep reactive ion etching (DRIE) was achieved.
  - A novel packaging method using LTCC substrate was proposed and demonstrated.
  - The nano capacitive gap width with smooth surfaces and less damage was obtained by utilizing neutral beam etching technology.
  - Narrow capacitive gaps as well as a capacitive silicon resonator structure were successfully formed by metal-assisted chemical etching.
- Silicon resonators were hermetically packaged on the basis of the anodic bonding technique. First, the structures of the resonator were transferred onto the LTCC substrate using anodic bonding of silicon and LTCC for electrical interconnections. Then, the resonator structures were hermetically packaged by the second anodic bonding of silicon and Tempax glass for encapsulation. The measured resonant frequency of the packaged device is 20.24 MHz and a high Q factor of 50,600 was observed without any kind of amplification.
- An ultra-high Q factor and low motional resistance were demonstrated by a long-bar-type silicon resonator. The theoretical analysis and the experimental results showed that the long-bar-type resonator body has a higher

Q factor and lower motional resistance that those of a shorter resonator body. Additionally, the device was hermetically packaged using an LTCC substrate based on anodic bonding. The resonant characteristics before and after the packaging process were evaluated. The resonator was excited in the extensional mode at a resonant frequency of 9.69 MHz. The Q factor measured for this device was 368,000 at vacuum chamber pressure of 0.01 Pa and 341,000 after packaging.

- A fabrication method of silicon resonators using neutral beam etching (NBE) technology was proposed to obtain a narrow capacitive gap for small motional resistance. Frequency characteristics of the devices fabricated by NBE with the resonant frequency of 9.66 MHz with a length of 500 µm, width of 440 µm, and thickness of 5 µm were evaluated, and a high average Q factor value of around 78,000 was achieved. Additionally, the devices fabricated by both DRIE and NBE were evaluated and compared to each other. The devices fabricated by NBE showed that the motional resistances were reduced by almost 11 times and their output signals were increased by approximately 15 dB compared with those fabricated by DRIE. Especially, devices fabricated by NBE provided higher Q factors (from 75,000 to 82,000) than those of devices fabricated by DRIE (from 57,000 to 66,000) in the comparison of the same resonator parameters and measurement conditions. The resonator fabricated by NBE realized a higher Q factor, lower insertion loss, and smaller motional resistance than those obtained by DRIE. A new approach to fabricate silicon micromechanical resonator using NBE was proposed.

- Metal-assisted chemical etching (MACE) was investigated for the etching of patterns of different sizes. Large patterning areas as well as a combination of large and narrow patterning areas were successfully formed. A capacitive silicon resonator with resonant peak of 81.4 MHz, quality factor of 4000, and motional resistance of 89 KΩ was fabricated by MACE. This proposal exhibits a potential way for the simple process and low-cost fabrication of micro/nano devices.

- Mechanically coupled capacitive nanomechanical silicon resonators to detect the small motional capacitance for attaining a low motional resistance toward emerging sensing, image, and data processing technologies were designed, fabricated and evaluated. Resonant peaks could be observed, which shows that most nanomechanical resonators are mechanically coupled and synchronized.

- High-order mode capacitive silicon resonators with the driving electrodes along the resonant body were produced and examined. It was demonstrated that high-order mode resonators could achieve lower insertion loss and smaller motional resistance than those of low-order mode resonators. Fixed–fixed beam capacitive silicon resonators, as well as other types of capacitive silicon resonators including bar type, disk type, and square type, could be employed for this proposal for high-order vibration modes with the low motional resistance.

- The capacitive silicon resonator with movable electrode structures can reduce the motional resistance for lower insertion loss and also increase the tuning frequency range to compensate for the temperature drifts of silicon resonators. Frequency characteristics of the silicon resonator of resonant frequency 9.65 MHz with a length of 500 μm, width of 440 μm, and thickness of 5 μm were evaluated, and a high Q factor of 49,000 was achieved at a polarization voltage of 25 V. The measurement results showed that the motional resistance is reduced by 200 times, the output signal (insertion loss) is increased by 21 dB, and the tuning characteristic of the frequency also increased by 7 times than that of a resonator without a movable electrode structure.
- Capacitive silicon resonators with piezoresistive beams (piezoresistive heat engine) were proposed and demonstrated. Capacitive silicon resonators with single and multiple piezoresistive beams were fabricated and evaluated. Improvements in the insertion loss and reductions in the motional resistance were achieved. The motional resistance of the proposal devices could be reduced by 90% in comparison to the conventional design. Additionally, the tuning frequency characteristic with the piezoresistive effect is increased by 165 times over that of the structure with only the capacitive effect.

The contents of this book are based on our experimental research, and we provided useful information with several examples including fabrication technologies and design considerations. It is our hope that this work may be a useful reference for those working in the field of micro/nanotechnology.

Index

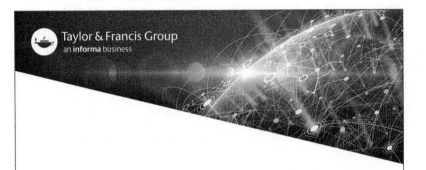